SERVING WITH HONOR

SERVING WITH HONOR

A SOLDIER'S LESSONS IN LEADERSHIP

By Colonel Glenn Schmick

PREFACE

AFTER HEARING ABOUT MY thirty-plus years of stories of the army, my son Hunter told me I should write them down. I initially laughed, but then he asked me again to write them down when he was a junior in high school as a graduation gift to him from high school, and it became a catalyst for this journey. I failed miserably at completing it in a year. In fact, by the time of this release, Hunter is a sophomore at Auburn University in Army ROTC and Business and is well on his way to becoming the next generation of our family to serve. The stories I tell are true, even if they seem hard to imagine, and only the names of those who were less than stellar have been changed. This journey was healing and, I hope, was a method to share hard-learned life lessons and a way to honor so many amazing soldiers whom I got to serve with and learn from. The army is unlike anything you can experience in the civilian world. It is hard, rewarding, exhausting, inspiring, and challenging—because it has to be.

Special Thanks to Dr. Corbin Williamson, Dr. John Geis, Dr. Ashly Adam Townsen, and Dr. Elizabeth Woodworth for their mentorship, guidance, and inspiration in my literary journey.

CHAPTER 1:
JOHN WAYNE AND THE FIRST SERGEANT (1986–1990)

❦

True Leaders Inspire Others by both example and action.

JOHN WAYNE AND THE first sergeant—two icons that elicit strong feelings from just about every soldier who has ever served. However, for my particular army journey, I was mesmerized by these two powerful forces long before I ever donned a uniform. As a latchkey kid from a divorced family, I was on the vanguard of the wave of children in the 1980s and beyond who found themselves without a traditional family setting. At the same time, we were from an age without a constant digital connection. We watched the news in the evening, played our cassette tapes in the morning, and hung out outside until way past dark.

The Cold War existed, but the concept of nuclear war and a communist takeover of the world failed to elicit fear in my generation. These threats seemed too distant to resonate in our daily lives or even compete with the feats of Michael Jordan or the birth of MTV. War and the military appeared only in short bursts in our collective consciousness. The failed Iran hostage rescue, the Beirut bombing, Grenada, and even Panama would flash on the nightly news one night, only to be quickly replaced the following news cycle. These conflicts did not produce any national heroes with which my generation could identify or inspire us to look toward the military. The shadow of Vietnam specifically impacted our exposure to the military. There was no Patton, Eisenhower, Audie Murphey, or other living military legend that I can recall from my childhood seen in daily life.

Instead, my earliest perception of the military was cast by Hollywood. From third through ninth grades, I lived in three states and five homes. This, coupled with my pronounced stuttering, shaped me toward introversion. As a result, I developed three habits: a love of wandering through the woods alone, and when at home, a love of reading and watching TV. Of course, in those days TV consisted of fewer than five channels at any given time based on your skill in positioning the rabbit ears (antenna). It was in this context that an introverted boy discovered his first role model, John Wayne.

I would be stretching the truth if I said I remember which movie I saw first. But I can today clearly remember the feelings these movie scenes awoke in me. At the time I could not tell you if the events portrayed were based on fiction or fact, and at the time I sure could not tell you a list of character traits displayed by John Wayne. However, whether I understood it or not, he was laying the foundation of the values that I would be drawn to throughout my adult life. John Wayne, born Marion Robert Morrison, starred in over 140 movies that any child of the 1980s could see at least two of on any given weekend on one of the five channels available.

I was mesmerized by his leadership, by his unwavering conviction to live up to something greater than personal reward. Whether it was his steadfast determination in the face of impossible odds in *The Alamo* or *She Wore a Yellow Ribbon* or his calm lead-from-the-front style in *The Longest Day* or *The Green Berets*, each cinematic production was motivation and mentorship to me as a child. His words were as good as gold, his values were never shaken, and his resolve was unquestionable.

As my home life devolved into violence and fear from an alcoholic father, a lack of stability from constant moving, and my lack of confidence brought on by stuttering, I found John Wayne gave me a positive image of what right must look like. The Duke, as he was sometimes called, was my first superhero. Years later when I became a father, I hung a tin picture in my son's room of the Duke with a quote that said, "Courage is being scared to death and saddling up anyway," proof that time had not diminished his impact on me. I never met John Wayne before he passed, but in 1985 I met a man who was the embodiment of all those character traits the Duke embodied in his movies.

In the spring of 1985, my brother and I were living with my father in Maryland. My father's second marriage of just over three years disintegrated almost overnight, taking with it our income, home, and short-lived stability. To restart, we moved for the sixth time in about a decade, this time to North Carolina for my father to take a management position in the US Postal Service. To complicate matters, my father's new girlfriend, whom we never met, moved in a few months later with her eight-year-old son. My brother, who was a rising junior, and I, a new high school freshman, prepared to enter a new school in a new state dirt poor, unstable, and struggling for hope and guidance.

It was in this troubled environment that we went to register for school, and God intervened without us even recognizing it. While we were registering, the school guidance counselor mentioned Junior Reserve Officer Training Corps (JROTC) as a possible class elective. My brother and I had never heard of this program since it was not popular in Maryland, and we knew nothing about it. The counselor in a few short words described how it was based on the army and taught map reading and marching, and you wore a uniform. She might have said more, but my mind was already formulating an image of John Wayne, and it sure sounded better than home economics.

Soon it was the first day of school, and it was on this day I met the most influential man of my life. Most of the first day had been a cauldron of stress and anxiety. I was a barely-five-foot chubby boy thrust into what then was a massive school of almost 1,500 students, with zero friends, no style, and no remarkable traits not counting my shyness. The high school students around me appeared as giants or movie actors. I was completely intimidated. I didn't think I would survive that day, and I was convinced that if by some miracle I survived the first day, I would never come back. My perception was unshakable until I entered the JROTC room for the first time.

As I entered the JROTC room, which was a double-wide trailer behind the school, I saw a man standing at the front of the class addressing each student as they entered the class. He would make a comment about our hair or clothes, anything to get a response from us and break up the tension of the first day. He said his name was First Sergeant Donald L. Chumley, and we would call him First Sergeant (First Sergeant). He was not an impressive-looking man in stature. He stood five six or five seven and didn't tip the scales above 140 pounds by my guess, and he looked old to my young eyes. I knew he had to be around the same age as my father, but the lines on his face were deep. However, most impressive was his presence. It was commanding. His eyes were laser focused, his voice boomed with confidence, and his persona was engaging.

Author as an awkward fourteen-year-old

Over the next four years, First Sergeant Chumley changed the course of my life. I started my freshman year with Ds and Fs as I struggled with my home life. My attitude was almost at the point of giving up on becoming *anything.* However, First Sergeant refused to accept defeatism in any of us as students. During my freshman year, he said, "I don't care what your grades are, but you will not be disrespectful in school. Shut your mouth and listen to your teachers—period." It was such a shocking statement from an adult that I was not prepared to hear it, but it flipped a switch inside my head. I found purpose in his direction; his words were simple, direct, and meant for me—a call to action. That year was the last time I ever got a bad grade. In fact, in my junior and senior years, I made almost straight As, allowing me to graduate in the top 5 percent of my class.

But why did First Sergeant have such a profound impact on me? I believe it was because he embodied so many character traits I had seen on the big screen in John Wayne, but in real life. However, the most powerful aspect I saw in him was his "lead-from-the-front" mentality. He never asked us to do what he would not do himself. This character trait and his model became my metric for measuring leaders, especially noncommissioned officers (NCOs), over the next thirty years.

First Sergeant was such a hardworking teacher, it was easy to overlook his history. I didn't know then that the first sergeant of any unit was the senior NCO charged with the care and welfare of the entire unit. I learned he was an infantryman, but I did not understand the concept of different jobs in the army. I was not savvy on how to read army uniforms back then and didn't readily notice his two Purple Heart medals denoting injuries sustained in combat or his multiple Bronze Stars for valor. I also didn't understand that the small marks on his sleeves denoted on one side his more than twenty years of service and on the other side months (totaling years) spent in combat.

Here was a man who was a combat infantryman with multiple vicious tours in Vietnam who filled the vending machines for the school, picked up trash when he saw it needed to be done, drove a bus to get JROTC students home, and helped the school any way he could. He didn't demand special attention, he didn't ever say it was not his job, and he never walked by a problem without making a difference. In many ways he was like Clark Kent—a regular guy working hard and easily overlooked.

But like Clark Kent's hidden personality of Superman, First Sergeant Chumley had, just below the surface, an alter ego that hinted at the resolve and steal of the man and warrior. First Sergeant was a smoker and often would smoke between classes. But instead of flicking cigarettes or tossing them half used, he would simply turn them over in his fingers and put them out with his thumb, then slide them back inside his cigarette pack for another time. When teaching us first aid, he would pick the largest student, well beyond two hundred pounds, and lift him from the ground and carry him in the fireman's carry. He seemed to simply will his body to do what he needed it to do.

Two events showing his resolve stick out in my mind even after all these years. First, there was a day when a group of gang members came to the JROTC trailer looking for one of our students to attack him. The student was not a good student, nor did he contribute much to the class. However, that did not matter to First Sergeant Chumley. This student was his student. The leader of the gang was six two, easily 240 pounds, and came in talking aggressively and said he was going to beat the student to death. Our JROTC officer in charge didn't move from his desk, and the rest of us froze in shock and fear. However, First Sergeant simply walked over to the group and told them to leave. The leader began to talk a lot of trash about who they were and what they were going to do.

First Sergeant stepped directly up to the leader with his nose inches away from the thug's chest and very plainly said, "Make your move." It was not what he said but how he said it that mattered. He did not raise his voice, increase his tempo, or make any motions or gestures. It was as if he knew there might be pain and blood, and that was just fine. He did not utter another word. His eyes locked onto the leader, and in an instant, his icy stare took all the fight out of the would-be assailants. The leader stepped back, raised his hands, and said, "I don't want none of you, First Sergeant." And just like that he took steps back, turned, and left. Thinking back now, I could feel the tension in the room, and I remember thinking, *What kind of man can face such danger so calmly?*

The second incident occurred on a trip to Disney World. To set this story up, you must understand First Sergeant was always finding ways to take us poor North Carolina kids on trips for little or no money. Rappelling, caving, winter survival, drill ceremony competitions, or trips to visit Washington, DC. He would use his military connections to find deals or stay on installations, rent a bus, and then get hired as the driver (he was a licensed eighteen-wheeler trucker), and have us sell some M&Ms to offset other costs. Then we would work in squads to cook breakfast, make lunches, pitch tents at campgrounds, or clean a military barracks we would stay at to reduce costs. If you respected your teachers and behaved, you could go on the trips. He showed us there was so much more to life than where we were. I can honestly say I learned more drill, first aid, mountaineering, and field craft as a high school student than I learned in thirty years of army life.

However, this one specific trip was epic. First Sergeant took more than thirty of us all the way to Florida to see Disney, an alligator park, and the space center. We spent almost a week on a campground, having the time of our lives. I was a junior or senior cadet by then and worked closely with First Sergeant and his wife (another retired first sergeant), who would always come along on these types of trips and chaperone the girls and direct the chow (food) preparations. It was in my now-leadership role that I saw the mental toughness of First Sergeant our very first night at the campgrounds.

The bus ride down had been uneventful, and most of us had slept or chatted during the ten-plus-hour drive except First Sergeant, who quietly drove the whole way. Once we got to the campground, First Sergeant led us to unpack, set up camp, and start cooking dinner. Just after dinner, most of us settled in and prepared for the fun of the next day. I went to check in with First Sergeant and found him off to the side and out of sight. His pants were rolled up and his feet were soaking in a small plastic container. As I walked up, I was preparing to say something funny about his feet when I noticed they were covered with blisters from the ankle down. These were not rubbed friction spots; I could immediately tell they were burn blisters. The casual look on his face did not match the look of extreme pain resonating from below the water.

I immediately asked him what happened and if we needed to go to the hospital. It was obvious to me that a stove or something else around the camp had burned him severely, but I was wrong. First Sergeant told me the bus had developed a serious defect almost immediately after we departed. The engine heat was directly entering the driver's compartment, and the intense heat burned his feet. I could not fathom what he told me. For over ten hours, he drove the bus with his feet cooking and not once mentioned it or even considered stopping the trip. Since this was before the days of cell phones, he knew any attempt to find a place to fix the bus would have ended the trip for us, and he would not entertain the idea. He told me he would be fine and not to tell the other cadets so they would not worry.

We spent the whole week in Florida. First Sergeant got the bus fixed and drove us everywhere. However, he never limped after that night and never showed any signs of injury. Again, he just seemed to will his body to deal with pain with intense mental focus. His wife once told me that after his time in Vietnam, it was not uncommon that when he ran and then bathed, shrapnel still in both legs would just work its way to the surface and come out in the tub. However, it didn't seem to faze him in the least. I knew he had to feel pain as he was not Superman, but I also realized he refused to let anything stop him once he was determined to do it. It was in him; I learned the power of the human spirit and the capability that lies within all of us.

It was the lessons First Sergeant taught us almost daily about being good citizens and people of honor and determination that profoundly changed the course of my life. As I neared graduation, I could not think of any other career path than to find more men like First Sergeant Chumley and be one part of them. I had found a place where I could be accepted, make a difference, and be tested. I joined the North Carolina Guard at seventeen as an 11B infantryman, but again First Sergeant told me to alter my path. He told me to apply to college and become an officer.

At first, I thought he was insulting me. I didn't know much about army officers because the only one I knew (our senior JROTC instructor, a lieutenant colonel) didn't seem to do much at all. Additionally, First Sergeant was always making fun of officers and stating that only soldiers and noncommissioned officers (NCOs) worked. I asked him why he told me to go to college and become a hated officer and not go straight into the army as a soldier after graduation. He said, "Because you have the potential to be one of the good ones."

I had potential? His unemotional and casual response was one of the most inspirational statements anyone has ever told me. At seventeen, I was now a product of John Wayne and First Sergeant.

CHAPTER 2:

FRENCH COMMANDOS AND POISON OAK (1990–1991)

❧

Perseverance and adaptation are critical
to the team and yourself.

I took First Sergeant's advice about college, but first I had to honor my commitment and head to basic training for the National Guard. I went to basic right after I graduated, before I turned eighteen. Looking back, I had to be a recruiter's dream. I scored a 97 on the Arms Services Vocational Aptitude Battery, was medically sound, and had never been in trouble with the law. I still remember the counselor at the military entrance processing station (MEPS) saying I qualified for any job, and there were several with bonuses. I told him I wanted infantry or nothing.

He said, "I will go see if we can find a slot." He came back and said he had just one slot left if I took it today.

Of course, I jumped at the offer, thinking how lucky I was to get the last slot. Looking back years later, I just laugh at it. I am sure there were dozens of slots, but he sure knew how to close the deal with me. I was so naive.

In June 1990, I stepped off a small commuter plane in Columbus, Georgia, in the stifling hundred-degree heat. I thought I knew what I was getting into. My brother was a paratrooper in the Eighty-Second Airborne, and I had had four years of JROTC, so I was nervous but confident at the same time. Besides, I had watched almost a hundred John Wayne movies, and I was sure they had shown me all the army secrets.

Boy, was I wrong. Basic training at Sand Hill, Fort Benning, was like walking through a soccer tournament with a Kick Me sign on your back. I hadn't gotten settled into a routine before I was kicked in the stomach, upended, and flung into another dizzying dilemma.

In just eight short weeks, I got hard, the hard way. I lost 20 percent of my body weight (thirty pounds) in a diet and exercise routine that was "medieval." We foot marched everywhere and every day. I got blisters on my feet from new boots, blisters on my hands from carrying my weapon, and blisters on my legs and back from my gear. At the same time, we were given only one minute to eat with a big spoon at each meal. And of course, we could only eat after downing two large glasses of water. The combination of constant motion and too few calories made us shed weight rapidly.

However, weight loss was only part of the infantry's crash course in converting our soft civilian tendencies. Our drill sergeants confiscated our deodorant on day one and restricted us to only two less-than-one-minute showers a week. In the June and July Georgia heat, we became so foul smelling, the chow hall (cafeteria) ladies who served us all wore clothespins on their noses. Our armpits and groins bled from the friction and filth while our minds reprocessed what pain was versus just discomfort. Over time we didn't notice how bad we smelled, and the filth and sores began to feel normal. Those who could not adjust quickly disappeared in the first few weeks.

I will never forget my basic training drill sergeants and commander. It was a true omen when we arrived and learned that our senior drill sergeant was named Tyranny and our commander was Captain Dye. My other two drill sergeants were, however, the most feared within my forty-eight-man training platoon. I will call them Ford and Stevens for our story. Drill Sergeant Ford had huge biceps, and when his sleeves were rolled up, his guns were on full display. Drill Sergeant Stevens was a wiry young drill who always seemed to be jacked on two or three cups of coffee at a time. Both made a huge impression on me, but neither of them finished basic training with us. They both got selected for French commando school.

I later learned that "French commando school selection" was what they told us privates when a drill sergeant was removed from his position for a violation. Of course, they did not want the appearance of their omnipotence to be dissipated, so French commando school it was.

In the army of the late eighties and early nineties, there was a slow change of the times of acceptable behavior. The official standard was that you could no longer cuss at privates or hit them. But these changes at the strategic level had not filtered down into the infantry culture yet. Instead, innovative soldiers developed workarounds. We were called DICKs (dedicated infantry commie killers), BITCHes (badass infantry-trained commie hunters), and many more acronyms designed around popular cuss words of the day. When it came to hitting us, the offending private would be used in a hip-pocket training event. "Hip pocket" was a term meaning an unplanned training event named after a small skill-level book designed to be carried in your hip pocket. As an example, the drill sergeant would have the soldier stand at attention and point out how to position his feet. While doing this he might kick the soldier's feet or punch his leg to highlight the reference. However, we all understood the soldier was getting a physical and mental scuffing.

Though some old-school techniques were falling out of favor, the ability to just smoke joe was alive and well. "Smoking joe" meant to just have privates continually do physical exercise. I can remember getting smoked for hours on end. Grass drills were a drill sergeant favorite, like what you may witness in high school football practice. The cadre would ensure to force hydrate us continually to possibly ensure safety, but we thought it was just to make more of us throw up more quickly. Our drill sergeants had complete power over our happiness and pain.

When I shot a 9 out of 40 on our basic rifle marksmanship test, I joined the Rocks for Rocks Club. The fact that I had not received my glasses yet did not dissuade my drill sergeants from seizing this opportunity to scuff me up. I had to run around the company, which was made up of about 160 men spread out over a quarter mile, as we marched back with my weapon above my head shouting "I ain't nothing but a dumb fuck" with the other dozen or so poor shooters. When we returned to the barracks, we had to go get a rock and tape it to the butt of our weapon so it would dig into our cheek. Then, while we held the weapon against our cheek, Drill Sergeant Ford would come by and punch us in the butt of the weapon, causing the rock to smash into our cheek and bruise our face. We were told to place the rock on the bruise every time we fired to ensure we had the same weapon position. The next time I went to the range, I shot a 40 out of 40, so I guess it worked. Or maybe it was the fact my glasses had arrived?

Drill Sergeant Ford might have been the most physical, but it was Stevens who was the first to get selected for French commando school one sunny Sunday. The entire company, about 160 privates, was sitting under our barracks on the cement cleaning weapons. We were in the newly created "starship barracks," which were built on columns to allow for a large outdoor space under them for soldiers to train and conduct business, so it was home to us. On Sundays, normally only a few drill sergeants managed us unless we were in the field or at a range. These drill sergeants would keep us busy with hours of barracks cleaning, weapons drill, or other mundane tasks.

This Sunday Drill Sergeant Stevens had the duty. I think we were about two or three weeks into basic, and by then we had about four or five privates who had decided the infantry sucked, and they wanted out. We had a few on suicide watch. I believe by design; these privates were openly shamed to ensure more of us would not look for an easy way out. They would have their shoelaces removed, forced to have two other soldiers go with them everywhere, including the bathroom, and even make other soldiers sit beside their beds while they slept to ensure the depressed soldier would not hurt himself. But they almost never went away or left the army quickly or easily.

On this Sunday, one of our suicidal privates had decided he had had enough. He went into a closet and drank over a quart of liquid floor wax. He then immediately told the other privates what he had done. Drill Sergeant Stevens became unhinged. He began to scream at the private and tell him to go over to the orderly room (basically the front office of the company). The soldier began to walk too slowly toward the room for Steven's liking, and the drill sergeant began to slap the back of the private's head repeatedly in front of the entire company. It didn't take long for the troop to collapse on the floor, curl up in a ball, and begin crying. Another drill sergeant came out of the orderly room and got between Drill Sergeant Stevens and the now-fetal private. The new drill sergeant yelled at two other privates to grab the suicidal trooper and throw him into Drill Sergeant Stevens's truck so he could take him to the hospital. Stevens responded, "Fuck that. I don't want this piece of shit in my truck. Throw him in the back of the deuce and half"—an army open-backed transport truck. And that is what they did. The next day Drill Sergeant Stevens was off to French commando school.

This event left a huge impression on me. I played back this event in my mind again and again. At first as a young trooper, I was torn between being glad it wasn't me and empathy for the young trooper and disgust at his weakness. Then I started thinking about the drill sergeant; was he wrong to assault him? And what could he have done differently? He was young and passionate. Was this a good thing, or was it bad? I finally realized the army lived on the edge of society norms and the violent character of war, and soldiers and leaders were always torn between these two worlds. My mind flashed back to high school English class; I was felt like I was reading *Lord of the Flies*.

It was only a few weeks later that Drill Sergeant Ford followed Stevens to French commando school. We were once again in the common area under our barracks as a company. This time we were rotating between small training stations practicing for our basic infantry skills testing. We were quizzed on rocket launchers, map reading, first aid, machine guns, and explosives over and over with the expectation we would memorize every specific detail of our craft. Drill Sergeant Ford was running the station covering the use of hand grenades. The station consisted of five or six grenades sitting on a large table.

Two soldiers would step up to the table at a time while the rest of the troops faced the other direction about twenty feet away. Each soldier would be asked a type of grenade, its characteristics and uses. Drill Sergeant Ford was intimidating when he asked questions and would smoke a joe when they got it wrong. I remember sweating when I stepped up, and I was keenly aware that I hadn't taken a breath until I answered his questions and was told, "Move out, DICK"—meaning I did good!

Unlike the suicidal soldier incident, I did not see Drill Sergeant Ford's actions, but I sure heard it. It was about thirty or forty minutes after I finished at his station. There was a loud crash that cut through the huge noise of over a hundred privates testing. The common area went quiet except for some moaning and Drill Sergeant Ford cussing. Someone called out for a medic, and at the same time, the other drill sergeants formed us up and marched us away from the area. However, before the march was complete, word spread up and down the formations of what had occurred. Two soldiers had failed to answer their grenade questions several times in a row, and Ford got extremely mad.

According to the story, Drill Sergeant Ford grabbed the huge heavy table in a fit of rage and flipped it over. Unfortunately, the table's edge caught both privates across their feet. They went down like sacks of flour tossed off a truck. Later that day both soldiers returned to the unit, each with a foot in a cast. Rumor spread that night that at least one of the troopers' feet was not bad originally after the incident, but he had further damaged it at the request of the rest of his squad to ensure the drill sergeant got in trouble. Whether it worked is hard to say, but we were told the next day that Drill Sergeant Ford had been selected as well to attend French commando school. I thought this school must be tough, because they seemed to select the angriest and hardest NCOs to go.

If my experiences in basic training formed one opinion of the army, my experience in advanced individual training shaped another. I was a split option enlistee, which meant I did my basic training one summer and came back the following summer to complete advanced individual training (AIT). This time when I arrived at Fort Benning, the drill sergeants took on a mentorship role and treated us much better. It was here I met who I will call Drill Sergeants Jones and Craft. Jones was a mountain of a man with a quiet air of confidence around him. He stood well over six feet and 260 pounds. His clean, black, shaven head looked custom fit into his drill sergeant hat. I cannot recall him ever raising his voice much, but he didn't have to because when he spoke, there was value in his words. Drill Sergeant Craft was a great complement to Drill Sergeant Jones. He was an old (in his thirties) and slightly heavy man who was nearing the end of his career. He seemed to choose his words carefully, as if trying to decide which pearls of wisdom were right for each occasion.

AIT was one of the best times of my life up to then. As trainees we had moved beyond just breaking into the army, and now we were starting to learn our field craft. Patrolling in the woods, conducting training maneuvers with live ammunition, and functioning as a unit with success helped us begin to understand why basic training had been so critical. We were now conditioned to focus on the needs of the unit above ourselves. We were no longer city boys, Hispanic, white, or anything else other than infantrymen. I found myself the platoon guide after a few weeks, which was the trainee platoon leader, and I was having a great time until the last field exercise of AIT. It was then that I got kicked in the stomach twice.

It was July 1991, and the Georgia woods were smoldering as we road marched before dawn into the woods that would be our defensive position for the next week. I was feeling supremely confident in both my skills and my platoon's. Before 1000 hours we arrived at our location and began to occupy the defensive position. I positioned the squads along a small slope forming from 4 to 8 on a clock face. The remaining two platoons tied into us on either end, completing a 360-degree defensive perimeter. Drill Sergeant Craft told me to dig a platoon headquarter position behind the line of foxholes the third platoon had begun to emplace.

I scanned the terrain and found, near the center of the line, a large tree that was surrounded by thick green leaves and weeds at the base. I thought this would give me both excellent cover from the enemy and the sun. I dragged my gear over and began to clear some of the foliage and roots before I paused to eat lunch. The rest of the day passed in the traditional infantry way—we dug while others provided security, and then we switched. I would go up and down the line about once an hour to check on everyone and ensure they were drinking water.

We were told we would begin patrolling at first light, so I hit the trench before dark. For those who have never used the great outdoors to relieve your bowels with a hundred of your closest friends around, hitting the trench required you to go over to a trench dug by troopers and fold your e-tool (army small shovel) so you could tuck it under one cheek of your butt to balance while you attempt to go. It required both balance and a lack of shyness because the trench was within the perimeter, to keep you safe. The unintended side effect was you could be seen by dozens of other troopers at any one time. I quickly finished my business and settled in for the night. I slept straight on the ground and fought with chiggers, mosquitos, and a few ticks to earn my peace and quiet.

The next morning, I volunteered to lead the training patrol under the watchful eye of DS Jones. It was lightly raining as we moved out, and the woods were silent. I thought I was doing well until DS Jones stopped the patrol and told me to quit weaving through all the thick brush. I found out later that my path kept his six-foot-plus frame in constant contact with tree limbs, and his drill sergeant hat was knocked off his head every other step. The patrol ended without incident as the rain let up.

As we got back to our fighting positions, my legs began to itch. I knew the chiggers and mosquitos had gotten in a few licks the night before, but this was different. I could feel heat coming off my legs, and I thought I'd better check. When I dropped my pants, I knew I was in trouble, as my legs above my boots were swollen to almost twice their normal size. I went to the medic, and he said it looked like poison oak. I had never seen poison oak before, but when I was young, I suffered a lot from poison ivy until I built up a tolerance for it. I went back to my position and covered my legs with pink calamine lotion, lime-green Chigg Away, and a top layer of bug spray.

That night was horribly painful for me. I dug, scratched, and tore at my legs searching for relief. I don't remember much of the next few days because I was focused on the misery. By the fifth day, I was exhausted with almost no sleep. Drill Sergeant Cage noticed my pained look in the chow line and asked what was wrong. I explained my ailment, and while I was talking, he looked over toward the perimeter. He asked where my fighting position had been, and I pointed to the large tree. He shook his head, looked pitifully at me, and said that the tree was covered in poison oak. It suddenly made stupid sense. I had cleared the poison with my shovel, sat on my shovel to use the bathroom, and then patrolled in the gentle rain. I could not have helped it spread better if I had wanted to. To top it off, I kept sleeping in it every night.

Drill Sergeant Cage took one look at my bloody legs, now caked with green and pink crust from the Chigg Away and calamine mixing with the open sores, and told me I was going into sick call. Sick call, for those who are not familiar, is the time and place where the army puts medical folks to see soldiers as patients. Sometimes it is at a hospital or clinic, and sometimes it is in the field. In basic training, it was at the hospital on the weekends, so I was heading out of the woods. Normally, going to sick call as a young trooper in those days was a sign of being a wimp. However, I was so tired and miserable I did not care who thought I was weak. I now had poison on *everything* below my waist.

The company supply sergeant drove me back to the company area in the deuce and a half (2.5-ton cargo truck). She told me to give my weapon to the armorer and catch the bus shuttle to the hospital. She said under no circumstances was I to take my weapon to the hospital. She then drove off, and I headed to the arms room. However, when I got to the company area, the entire place was a ghost town with nobody around. I had only five minutes before the scheduled arrival of the shuttle, so I dashed upstairs and locked my weapon in my wall locker to ensure it wouldn't get lost. Then I sprinted to the bus, barely making it in time. When I got to the hospital, I was utterly lost in the maze of corridors.

I finally made it to the ER and was seen shortly by a nice, young female doctor. She saw the extent of my poison and told me she was going to give me a cortisone shot. After the shot, she said that I needed to sit there on the hospital gurney for a minute to ensure I wouldn't have any reaction. That was the last thing I remembered for about twenty minutes. When I regained my wits, I heard her saying how sad it was that a little shot put me down. I then saw a combat medic NCO dress her down. He said it was obvious by my appearance and smell that I had been in the field with little or no sleep. This NCO sticking up for me snapped me back to reality. I thanked the medical folks and headed back to my barracks; however, the whole way there I felt like I was drugged or drunk. I had reached my breaking point.

When I arrived back at the barracks, the lock was missing from my locker, and in its place was a government lock. It foreshadowed the second punch in the gut I was going to receive in less than a week. I went down to the CQ desk, frantically trying to find out where my weapon was and what had happened while I was at the hospital. The NCO at the staff duty said my lock was cut to retrieve the weapon and secure it in the arms room and that I was to report at 0600 the next morning. By then I had nothing left inside me, and I crashed on my bunk in sheer exhaustion. I awoke the next morning believing I had just had a bad dream until I saw the government lock was still on my locker.

That morning I stood at attention in front of both drill sergeants, Cage and Jones. I attempted to explain the simple logic in my actions to them, assured that they would see the right in my approach, and all would be forgiven. Instead, they cut me off and said my logic was irrelevant and I had broken an army regulation. The commander, I was informed, was considering UCMJ on me. UCMJ is short for the Uniform Code of Military Justice, which is the power a commander is given to act as a prosecutor and judge over his soldiers. I could not believe that one week ago I was the platoon guide riding high, and now I was physically a mess and facing criminal charges.

For the rest of the day, I did small work details around the barracks until the remainder of the company returned from the field that night. DSs Cage and Jones brought me back that evening and told me they had talked to the commander and shared with him their perception of me. He had decided to not UCMJ me, but I had to be punished. So it was agreed that I would be removed from platoon guide, and I would pull fire guard (a detail where one soldier is always on guard) for an hour every night until graduation one week away. There in that small room, I saw the role of a commander and NCOs so crystal clear that this event stuck with me my entire career—there is an art to balancing the needs of the unit and the care of soldiers.

In the end, army basic training and AIT fundamentally changed me, I believe, for the better. I learned how to persevere, to be a member of a team, and to adapt to anything. Looking back, this same experience shaped how I would view young soldiers as I became a senior officer. A part of me always tried to view my actions through the eyes of eighteen-year-old PFC Schmick. Learning about French commandos and poison oak was one of my greatest adventures in life.

CHAPTER 3

THUNDERSTORMS AND APPALACHIAN MOUNTAINS (1990–1994)

As a leader you always impact others—good or bad.

BASIC AND AIT CEMENTED in my mind that the army was going to be my profession. However, it was the people I began to meet in college ROTC that made me realize the army was going to be my way of life. I could probably write a book on just these amazing people I met at Appalachian State University from 1990 to 1994. Most of these folks are still my friends today, almost thirty years later. Most left the service long before the writing of this book, but the bond we shared carries through the decades.

When I try to figure out what the connection was, I always come back to the thought that it was the first time I was exposed for a long time to individuals who were all drawn to something greater than self. Even writing this sounds a little self-serving, but I am not trying to be. If you were to put us together in any other environment, I am sure we would never have talked much to one another because of our different backgrounds, styles, and attitudes. But in the military, all this bias had to take a back seat to completing something greater than oneself.

Up to this point in my short army career, I had been exposed only to good or great leaders. Even those French commando drill sergeants to me at the time were not bad; they were just overzealous. This changed when I began to serve with my National Guard infantry unit. However, before I begin, it is important to remind readers that the National Guard of today is not the National Guard of the 1990s. Today's units have numerous combat tours in Iraq and Afghanistan, and their veterans have greatly enhanced the service. However, as a new private first class in a mechanized infantry detachment in 1990, I found myself surrounded by superior officers and NCOs who showed me that not all leaders are good, and it was important to any organization to root out those leaders because their damage was infinite.

I can summarize my bad leadership experience in just two weekend drills at Fort Bragg, North Carolina. The first weekend we had drawn out our M113s (tracked troop carriers) for weekend weapons training. I loaded all the equipment onto the vehicle and realized there was literally no room to move within the vehicle. My squad leader had us load a "squad box" as large and heavy as a coffin into the back of the crew area, which nearly gave me a hernia. He made me both the M60 machine gun assistant gunner and Dragon anti-tank missile gunner, which meant my 150-pound frame was lugging an additional sixty pounds of new toys on top of my already heavy infantry kit. However, as the new guy I expected this treatment, and I didn't want to disappoint my squad leader. Since I was new to the unit, I wanted to make sure I made a good impression.

The first morning after we made it to the field, my squad leader told me that right after breakfast, I was to go into the post with another private to get our military IDs made at the ID center. So off I went and got my ID made by 0900; I still have that ID today showing a boy trying to look tough. I thought the day was going great, but my good luck seemed to end at the conclusion of getting my picture made. The other private and I stood and then sat outside the ID center for several hours. Cell phones didn't exist, and we had no idea where our unit was, so we just sat there in the North Carolina blazing sun. At about 1500 hours (3 p.m.), our first sergeant (the senior NCO of the company) came by in a Humvee (our basic field utility vehicle). He asked us how we were doing, and we told him we had been there for over six hours waiting on a ride with no food or water. He allowed us each to grab a pudding pack he had and told us a ride would be there shortly.

It was hard to hide how I felt in my national Guard ID photo

To a civilian, *shortly* may mean fifteen or twenty minutes. However, in the army it could mean fifteen minutes or several hours. On this occasion it meant several hours. When a truck arrived to pick us up, the sun was going down. We were relieved to get on the truck just before 1900 hours (7 p.m.). By the time I made it back to my M113, I had been gone for almost fourteen hours. I was dehydrated, starving, and just a little bit pissed off about wasting the day away. However, my irritation turned into disdain in the next few minutes.

My squad leader was not at all curious about where I had been or what I had been doing all day. I told him anyway, highlighting the fact that I had not eaten or drank almost all day. He looked over at me and said, "You missed dinner as well. It was served about two hours ago." My heart sank, but before I could respond, he stated, "I have some hot dogs, chips, and sodas in the squad box."

Yes! I thought. Finally I could eat. I jumped to my feet and headed toward the box, then he stopped me in my tracks when he said, "Each item is a dollar." I was stunned. My NCO, who was supposed to look out for me, not only hadn't checked on me and hadn't ensured I got food, now wanted to charge me to eat after I had carried his food in the squad box to begin with yesterday. I turned around, resigned to go to bed hungry.

It was at that moment that I learned what a bad NCO looked like in the army. I didn't expect all my leaders to be the fastest or smartest, but I did expect them to care. I tried to picture John Wayne doing what this squad leader had just done, and I knew he would have never done that to one of his men. From that day forward I evaluated NCOs in large part by how they cared for their soldiers.

In the following drill at Fort Bragg, I learned officers were not immune to not caring enough. Up until this weekend, I had not seen my lieutenant (the officer in charge of my platoon) for more than five minutes. He was easy to spot because he was easily fifty pounds overweight. However, this weekend he joined us as we hit the field without our M113s. We were told a serious thunderstorm was rolling in and we had better get set. Back then the E6s and higher leaders slept somewhere else, but I am not sure where.

The five of us who constituted the squad's E5s and below started reinforcing our ponchos above our heads and digging trenches around our sleeping bags. After about an hour we were in pretty good shape. We had selected high ground to avoid getting soaked on the bottom but had not gone so high as to expose ourselves to lightning. As the sun faded behind the skyline, we sat down to eat our MREs (meals ready to eat). We could see the lightning arching across the sky, and the wind picked up, rolling thunder to our ears. It was then that I saw our lieutenant coming to check on us—or so I thought.

Instead of inspecting our handiwork, he simply said, "You guys are set up near a red-cockaded woodpecker tree, and you must move." He shined his flashlight in the dying light at a tree fifty feet away, which had three painted bands around its trunk. This was the standard marking to show the endangered bird's habitat.

I was thinking that he had to be joking. There were no birds coming out in this approaching storm, and we were about to get hit hard by mother nature. But he said he was not joking, and we had to move at least a hundred feet away. Then, just like that, he was gone.

We ripped down our ponchos, grabbed our kits, and dragged our sleeping bags as fast as we could. All the light was gone by now, along with any hope we had of finding good ground again. We focused on just trying to find anywhere to tie five ponchos to a tree to create some type of overhead cover. I tied two corners to a tree shared with another trooper. However, the final two corners I had to tie to a small bush, which I was not happy about at all. But the storm was on us, with strong wind gusts and constant lightning and thunder. We would not get a chance to dig any ditches before the tidal wave of rain echoing through the woods rushed toward us. Instead, we jumped under our ponchos and attempted to dig from the inside to at least make a runoff trench.

The storm was a monster. I was scared but resigned to the fact that I had nowhere else to go. I crawled into my sleeping bag and zipped it up to block some of the rain blowing sideways. I fell asleep for a few minutes, I believe. However, I was awoken by something nudging me gently from above. I poked my head out and gazed up at my poncho spread above me. There I saw an engorged cow udder where the head opening of the poncho should be. In our hasty move, I had not goose-necked (tied a knot) in the head opening, and now it was full of a gallon of water, causing it to hang down directly over me.

The trapped water in the middle of my poncho was compounding itself by the second, and I had to deal with it. The weight of the water was causing the sides of the poncho tied to the bush to lean in more. In turn, the great angle meant more water rushed to the middle. Water was now dripping into my sleeping bag. However, the portion of the poncho designed for your head had a large opening for your face. I had pulled the drawstrings tight, but there was still a three-inch hole near the top of the bubble where the water was coming from. I knew if I did nothing, I would get water on me through this hole, but if I attempted to push the water up and out, it might collapse the poncho altogether.

I decided I had to push the water out. I sat up in my sleeping bag with the lightning and thunder all around me. I planned to push the bubble straight up with one hand while supporting the weight it would create on the poncho with the other hand. It would have been a great plan if it had only worked. But as I pushed up rapidly, water sought the path of least resistance, which happened to be the three-inch hole directly above my sleeping bag. In a flash, one gallon of water splashed directly inside my sleeping bag. I was wet, cold, and defeated. Like a good infantryman, I cussed, zipped up my bag, and settled into my watery cocoon.

The next morning as I attempted to swim out of my sleeping bag, I thought about the leadership I had witnessed the night before. None of my leaders had checked the area we were bivouacked in during the entire day. Additionally, no leaders had given us guidance on where to set up or shared our hardships. Finally, when the lieutenant did come to us, he decided the requirement to move outweighed the risk of the pending storm.

He taught me three things: Give your intent or live with what you find; like the famous German Panzer commander Rommel, eat and sleep like your men; and always evaluate the risk versus return in any mission situation.

However, this weekend was not done teaching me about bad officers. I met my first field-grade officer outside of ROTC, and it was as memorable as the reason I came to see him. Sunday morning, after I swam out of my sleeping bag, we headed off to execute a grenade assault course. This was a large obstacle course where you ran from station to station negotiating various environments to get within range to throw a practice grenade at a target. There were trenches to crawl in, logs to climb over, and all sorts of structures to throw your grenade into, such as windows, bunkers, or foxholes.

I began my run at the course easily enough. I was in great shape, and I loved to be doing a challenging course and some real training. I had made it through a little over half of the course when I started having a pain in my stomach stretching down to my groin. By the time I crossed the finish line, I felt like someone had kicked me between my legs. I told my NCO, and by the time I walked over to the medics, stationed in the middle of the course, I could not stand up straight.

The medics told me to drop my pants. We were out in the middle of a field with hundreds of troopers all around us, and I did not feel like putting on a show since I was not wearing any underwear. Not wearing any underwear was a common practice in the infantry back then, and it may be the same today. However, the pain was significant, and I was willing to run around naked if it would go away. The medics didn't have any gloves, so one of them emptied a Ziplock bag and slid it over his hand. So there I stood in broad daylight with my pants around my ankles and a man with a Ziplock bag on his hand grabbing my jewels.

I could tell this was not a fun event for the medic either, as he quickly finished his check and said I had to go to the battalion aid station, which was a small medical team in the woods with us. I was driven over to the battalion aid station, where another medic performed the same check—this time with a rubber glove, thank goodness. I now felt like someone had hung a thirty-five-pound dumbbell off my testicles and it was pulling my stomach out the bottom of my body. He told me he thought I had a hernia, but he was not sure, so he was going to send me to the post hospital. I didn't know what a hernia was, but it didn't sound good.

Since it was Sunday, I was dropped off at the emergency room, where I joined dozens and dozens of folks waiting to be seen. After about two hours, I was led back to an examination room. The medic had gotten the details of my complaint earlier and told me the doctor would be in shortly. After just a few minutes, in flew an army major in a hurry. He looked at me with what I would describe as disgust and told me to take off my pants. I could barely bend over by now, and I winced in pain as I attempted to untie my boots. He glared at me and said, "Quit your whining, and get your pants off."

The eighteen-year-old in me wanted so badly to demonstrate the pain on his body, but the soldier in me made me bite my lip. Once I got my pants off, the less-than-pleasant doctor grabbed me like he was trying to crush a can. If I had not been so pissed at him, he would have seen a tear roll down my cheek. But I refused to give him the satisfaction, so again I bit my lip.

After what seemed like entirely too long for one man to crush the potential offspring of another, the doctor released his death grip. He stepped back and said, "You have epididymitis with a varicose seal."

I froze with fear and confusion and shouted, "That cannot be, sir. I have only ever been with one woman."

He did a double take at me and said, "No, you idiot. Your nuts are in a knot. That's what you get for not wearing underwear."

Well, I tell you I didn't know if I should be relieved or terrified. But he issued me some muscle relaxers and told me to prop my feet up for the rest of the day, and they would figure themselves out.

As I lounged under the North Carolina pines that afternoon with my feet propped up on my rucksack, I thought about my encounter with this major. Maybe he was having a bad day, or maybe his workload was overwhelming for a Sunday. I figured I would never know. But I realized our brief meeting had painted a very negative picture of senior officers. I had come to him in shameful pain, and he had called me a whiner and idiot. This one doctor's single interaction affected the way I trusted and treated doctors for the rest of my career. Additionally, I realized that as you become more senior, routine encounters for you are big deals for young soldiers.

Meanwhile, between my adventures with the National Guard, I enjoyed my time at college. Out of the millions of experiences I had there, one stands out in my development the most. However, like most lessons, it took a long time for me to grasp the knowledge given to me. It was probably because when I was in my twenties, everything was a competition, and I hated to lose. I mean I hated to lose at anything. Now, I realize this incredibly powerful drive to succeed can and will have negative consequences if you are not self-aware. Being the best is not always the best for you, and when you are trying to be the best at something, you often miss so much around you.

Wesley was a fellow cadet in my class, and she was very talented. She was extremely kind and compassionate, but she always sought knowledge and focused her efforts on very analytical thought. I enjoyed her company, and when we went to airborne school together in 1992 after our sophomore year with another amazing cadet, Melanie, we had one of the greatest adventures I had had to date. Jumping from planes, staying in roach-infested hotels on the beach during the weekend breaks, and basically just being college kids was a blast. Coming from poverty and never having really taken a vacation before, in a weird way airborne school was fun. Of course, it was challenging, but the fear of jumping and the daily physical tests were an adrenaline rush.

During this same time, Larry Bauguess became an influential figure in my army development. Larry was another cadet, just one year ahead of me. However, we had served in the same National Guard unit together when I was a freshman. Larry's professionalism and drive to be the best officer he could become was inspiring. I remember thinking I was a fast runner for my unit; back then I could run two miles in the twelve-minute range. However, Larry without bragging would just go out and run his hardest regardless of the rest of us. He always beat me, but he did not revel in it. Instead he just looped around and ran right back down the course to help others finish at their best.

Larry became a key leader in ROTC, and later Wesley became the senior cadet in my year group because of her tremendous skill. Both drove me to be better, but my competitive nature drove me to interpret their success as my failure. Looking back, it was my immaturity and childhood experiences that shaped these thoughts. Today, I realize this drive to never quit or lose has cost me many friendships over these years. I know it damaged my relationship with both Wesley and Larry, but they were better friends than me and put up with me.

As fate would have it, Larry and Wesley fell in love. One of my greatest honors was to be a member of the saber guard at their wedding. I remember one of my ROTC officers asking me when I was going to marry my girlfriend about the same time. I said with brash arrogance, "If the army wanted me to have a wife, they would issue me one."

He said, "Sit down. I am going to give you some free advice. When you leave the army after twenty years, they are going to give you a medal. I will even tell you what it is—a Meritorious Service Medal. Then they are going to say nice things about you, but for no more than two minutes. After that ceremony the army is going to forget you exist because the army is not a person. You can love the army, but it cannot love you back. It is designed to go on without you. So remember, all you will have is the family and friends you make and keep along the way."

This statement, coupled with Wesley and Larry's love, drove this point into my military DNA. Over the next thirty years, I would echo this sentiment in every command philosophy and vision I shared with my soldiers. Later in my career, I began to call it the long game versus the short game. The short game was the task of the day, and the long game was keeping great soldiers in the army. I now firmly believe the long game was so much more important than the short game. By the time I became a brigade commander, I had written a few articles on the subject, and I wish I had learned this lesson and practiced it earlier in my career. My time at Appalachian State University was one of the best times of my life. My peers and leaders showed me that kindness, personal drive, and balance were more important than just the skills and tasks the army measures daily. Honestly, I was not able to begin to appreciate the gifts they showed me until after 9/11. Until then I was just not ready to see beyond the daily fight.

Larry and Wesley went on after college to serve and had two amazing daughters. However, Larry was killed in May 2007 in Afghanistan leading his troops as he had always done. Wesley's strength and faith to this day humble me. I wear Larry's name on a bracelet on my wrist to, in a very small way, thank him. He was and is the very best of our nation.

CHAPTER 4

WHITE THREAD
IN A SHORT TAB
(1995)

*You can endure more than you think and
still be kind to others.*

1SG Chumley pinned me as a butter bar and then dropped me for pushups

IN 1994, THE WAR that would take Larry was still over half a decade away, and I was a newly commissioned "butter bar," the nickname given second lieutenants by soldiers. I had been branched as a military police (MP) officer after requesting it. I had decided to go into the military police because I wanted to see action, and they got to carry weapons both at home on post and in combat. Additionally, my time in the infantry had taught me there was a lot of just doing nothing when not in the field. And I so "unwisely deduced" there was not going to be any major conflict during my time in the army. America was into nation building and peace keeping, and my best chance to see action was to be part of the force that supported those missions.

I thought I had everything figured out. My college girlfriend and I would get married, and we would travel in the army while I went off every few years to save some far-away country from itself. However, shortly after I arrived at MP training, my girlfriend told me she was heading back to Florida from college and planned to live there her whole life. She was and is an amazing woman, and I am glad she made that call early before our life got more complicated. However, at the time it crushed me. I now had no close family, no one to care for, and only the army to focus on.

At the same time, I had applied to go to Ranger school. It had always been a major goal of mine for some reason. The army was changing Ranger school back to a light infantry school, meaning MPs and almost all other branches would not be able to go in a year. My instructors told me I would have to graduate number one from my MP basic course, volunteer for extra jobs in school, and pass the pretest to get one of the last three slots for MPs. I focused all my energy and did exactly what they told me to do. However, in late January 1995, during the last week of school, after finally winning a slot to Ranger school, I hurt my lower back competing for the German Expert Military Badge. I was in excruciating pain, but I was not about to miss Ranger school. What I did not know until living with the pain for eight years and an MRI in 2003 was that I had ruptured two disks in my back just ten days before I entered Ranger class 5-95 on 05 February 1995.

Ranger school. Just typing those words is a near-religious experience for me. That school broke me; that school made me.

I arrived at Fort Benning on 05 February clueless as to what was in store for me. There was no internet back then, and all I knew was a packing list and an address to arrive at. I showed up with all my gear, a 156-pound body, and the final words of my MP instructors echoing in my head: "You cannot recycle"—redo any part of the course if you fail—"because the MP Corps will not pay for it." Those words drove me when in the first seventy-two hours, my class was ruthlessly whittled down from three hundred to just over one hundred. The evil these men, these black hats (as the Ranger instructors, or RIs, were called) could so effortlessly unleash was devastating.

The first seventy-two hours was called RAP, or Ranger Assessment Phase, and it was a series of mental and physical tests designed to weed out those who just did not belong in the course. We barely slept in those first three days. Instead, we broke ice and swam to execute our water survival certification; tripped and fell doing land navigation at night without lights; or got rinsed off with a firehose and stood at attention in the thirty-degree weather at 1 a.m. until someone said those magical words, "I quit." I found myself begging for someone to quit. The pain would stop, if for only a second, if someone would just quit.

By the end of the third day, my company ceased to exist. Only eight of us from one hundred arrived at Camp Darby after a miserable twenty-mile forced march in the sleet and snow. The eight of us were folded into what was left of the other two companies, and we began the first of four phases of Ranger school with just over one hundred men. It was in this phase that I gave myself to the tab ("tab" is the nickname for the embroidered Ranger patch worn above the unit patch on your uniform if you are Ranger qualified).

I had nothing to go back to, nothing to distract my attention. I first realized it when the instructors stopped training one day and brought the class together. They told us four Rangers had just died. We all immediately thought we were at war. They then said it was in training, and we thought, man, those Ranger units train hard, and accidents happen. They then said no, they died of hypothermia in phase four of the same school we were in. Unlike what the news media said about Ranger school stopping training, we only paused for about twenty minutes, and then we were told to go back to work. This news "hit different" and affected many students hard. I don't know how many quit that week, but I heard more than a few say, "This tab isn't worth dying over."

The news had the opposite effect on me. I vowed not to quit; I wanted to be a part of this brotherhood even if it killed me. This switch in my mind was cemented in a primeval *Lord of the Flies* way later that evening when we conducted our survival training cooking. You see, part of the training at this point involved killing rabbits or chickens and learning to cook them in the woods. I was selected to kill the chicken for my forty-plus-man platoon. The technique was to hold the chicken by its feet and pet his head until it relaxed. Then, in one motion you were to step on its head and yank on its feet, breaking its neck.

As I prepared the chicken, it was dark, and cold rain was falling. There were about thirty-five of us cold, filthy Ranger students all huddled under a wood-and-tin pavilion deep in the woods. My wet boot slipped off the poor chicken's neck on my first attempt, and the RI lunged at me to grab the chicken and finish it. But I was determined. I quickly stepped hard the second time and pulled with all my might. I pulled so hard his head came clean off his body, and as I held on to his legs, the headless bird attempted to flap his wings and escape. This may sound horrific to the reader, but blood spurted out onto the platoon, and everyone went wild. It was a haunting scene looking back, but at the time it was a rite of passage into the world of kill or be killed.

After surviving the first phase of Ranger school, we headed off to the second, now extinct, desert phase. There my fingers cracked open due to dehydration and cold because of the temperature swinging thirty to forty degrees in just six to eight hours. During the days we would blister from being in the sun and then at night spoon (a phrase meaning lie against one another to stay warm) and shiver all night. At this point, it is important to explain that in Ranger school, you carried extra clothes and a sleeping bag, but you were not allowed to use them. If you were caught, it was an automatic expulsion. The idea was to train you to carry heavy loads but survive with almost nothing. However, most of all, the desert taught me "Be ready."

In Ranger school you rotated your job daily. Most days you were a team member doing some small part of the mission. Carrying the machine gun; being point man, medic, or compass man. On other days you might find yourself one of the leaders who were in graded positions like platoon leader. If you failed these, you got recycled. However, in the first phase, we learned how to know when the changes in leadership were coming. One set of leaders would plan the mission, and then they would change the leadership for the execution.

So I was relieved when for our first desert mission, after three days of more survival training, I was selected to be an M-60 machine gunner. The machine gun might have been heavy, but not as heavy as leading the mission. We were going to air assault into a desert valley at night and then attack to destroy a satellite station. I was further relieved when they did not change the leadership before we boarded the helicopters. I assumed that during the new phase, leaders would go through the whole mission. So to be honest, I spaced out and focused on not throwing up or freezing, as the Vietnam vet pilots flew those old Huey helicopters with the doors open in ways that must have been illegal.

I was jacked and excited when we touched down, but my rucksack got stuck in the frame of the helicopter, so I had to yank it hard to get out. When I did, I fell straight back into a thorn bush. I lay there bleeding from the dozen piercings in my back and softly cussed. One of the RIs heard me and said, "Ranger, come here." He was going to kick me when I was down.

He told me I was now the heavy weapons squad leader for the mission, and my mind went blank. I mean totally blank. A Ranger platoon consisted of three normal squads of about nine or ten men. Then there was the heavy weapons squad, which had the same number of men but carried the three heavy machine guns and usually operated as the base of fire for a mission. This meant it normally maneuvered and set up away from the other three squads. Unfortunately, I could not remember a single aspect of the objective we were attacking. I had no idea where we were on the map, and it was dark in a flat desert. I immediately wished I hadn't zoned out before the mission.

I quickly worked my way forward and found the newly assigned platoon leader and discovered we were already way behind target time. So without even a minute or two to get organized, we quickly moved out through the darkness. I am sure that to the Ranger instructors, we looked like a "football bat."

About an hour later, the column stopped, and the platoon leader called for key leaders. He said this was our ORP (objective rally point), or the last covered and concealed position before you enter your assault positions. But unlike normal protocol, he said we did not have time to conduct a leaders' recon or prepare equipment for the assault. We had to attack now.

I thought this was ridiculous. I had three M-60s covered in sand from the rotor wash of the helicopters, ammo dispersed throughout the formation for the march, and no night weapons optics ready. But the worst part was that I still didn't know where I was or what the objective was. However, the platoon leader did not bat an eye. He said, "I will give you two SAWs"—squad automatic weapons, a light machine gun—"to augment your M-60s." Then he pointed to my right and said, "You go there. The enemy will be to your left, and I will attack them from the left." Then in a flash, he was gone.

Years later, I found myself reflecting on that platoon leader. He was handed a mission set up for failure, but he refused to accept it. I found myself, at that moment, motivated by his determination. I led my squad several hundred meters until I could see the outline of a compound in the low ground to my left. I spread out my three medium and two light machine guns parallel to the target. We were a solid three to four hundred meters away, and I had excellent fields of fire. I thought this was going to be perfect until I looked to my left and saw our assault force approaching the target. Instead of coming from my left front, they were coming from my immediate left, which meant that to get to the target, they would have to walk right in front of my guns.

I grabbed the radio and attempted to reach the platoon leader— no luck. I had no choice. I took off like a madman running down through the dark two hundred meters until I reached the man closest to my line. It was dark, so I misjudged my stop and smacked right into him. I kept the momentum and pushed him to the left. "Shift left, you jackass. You are going to walk across my fields of fire," I whisper yelled.

The message got through, and the line began to shift away. I turned around and sprinted back to my gun line. By the time I got my bearings again, I could see the assault force was almost within a hundred meters of the objective. I didn't get any signal for firing, but I knew if I didn't fire soon, the assault force would be compromised.

I gave the command "Gun one fire," which was met with *kerchunk*. *Kerchunk* is the sound of a machine gun bolt slamming forward without firing a round. Our first gun had failed. I called for gun two. Another *kerchunk* followed immediately by one round from gun three before it failed. The sand had done its job.

Meanwhile, the first gunner had already done an immediate action drill and reported ready. I called gun one, and he ripped off ten to twelve blank rounds into the night. Oh, that sound was pure heaven to me. I then called for gun two, as I was attempting to "talk the guns." Talking the guns was alternating between guns to ensure complete coverage of the multiple targets without overheating the barrels. However, guns two and three would not get into action. I had only one medium and two light machine guns in play. I ran over to gun one and said, "No matter what gun I call—you fire." I then told every M-60 to take all ammo to gun one. For the next three minutes, gun one fired continuously, almost to the point of melting down.

Once the assault was complete, we linked up with the main force and moved to our new patrol base before dawn. I waited for the RI to counsel me and fail my mission. There was no way I passed if he was on that gun line during the attack. But to my surprise, he gave me a PASS. He told me during the darkness he got separated from my squad, so he moved with the assault force. He heard me calling each gun and alternating them. I could not believe my luck. I told him I had to run down to the assault force and shift them. He told me he knew, and if I had not, he would have failed me. "But," he said, "do not ever push me or call me a jackass again." I froze until he cracked a small smile. What were the odds that the one guy I ran into was my RI? Better to be lucky than good.

The mountains of Dahlonega, Georgia, and phase three of Ranger school showed me luck runs both ways in the army. We picked up almost fifty recycles from the class before us when we arrived. We had lost a dozen or so in the desert, so our ranks continued to change. It was early March, and the Appalachian mountains were cold. My body was breaking down by now, and my ruptured discs were joined by swollen knees and burning thighs in a chorus of daily pain. The mountains would have finished me if it were not for my Ranger buddy, Rulla. He taught me it only takes a second to change the outcome for another human.

The terrain of mountain phase became the devil. All I remember about the mountain phase was the pain of going up and down and then doing it again and again. We covered thousands of vertical feet daily in the dark, rain, and wind. By our last twenty-four hours on patrol, I had two T-shirts tied around my knees, which were swollen like grapefruits. I knew I had done well on my patrol a week earlier, so I just needed to endure the pain for a little longer. However, the Ranger gods were not going to let me off that easy.

On the last day, the platoon drew a rare daylight search-and-attack mission. This was welcome news, as they assigned a foreign officer who had failed a patrol earlier to be the platoon leader. I really didn't care who was in charge until I was told I would be the platoon leader's RTO, or radio man. This sucked. It meant I would add another twenty pounds to my rucksack, and unlike everyone else who would dump their gear on an attack, I had to wear mine for the PL to communicate. This was just what my knees needed.

As we set out on the mission, it became evident that we were way behind our schedule and had to move at a hard pace to cover the several miles to the objective. When we got close, the patrol stopped, and our platoon leader quickly assembled key leaders to go execute a "leaders' recon." Everyone on the recon except me dropped their rucksacks and prepared to move out. As the RTO, I had to keep my rucksack on.

Our recon element left the perimeter and headed straight up. I mean they ran up the side of a steep hilltop, which was at a forty-five degree angle for two hundred yards. By the time I caught up with them at the top overlooking the objective, I was hyperventilating and barely able to stand as my knees were locking up. I crashed to the ground in pain, only to hear they were done, and we were heading immediately back down the hill.

I felt the words "I quit" forming in my throat, and I might have said them, but everyone had already started running down the hill, and I had no one to listen to me cry out. So I looked down the hill at the platoon below and then at my legs, and I did the only thing I could at the time. I started to roll, fall, and flop down the hill any way I could. I imagine I looked like a comedy act or a drunken sailor to anyone watching. My rucksack was flopping over my head, the radio hand mic was wrapped around my body, and my weapon managed to snag on every tree limb possible.

When I finally hit the bottom and dragged my body inside the perimeter, I was met by a smiling RI who simply said, "That's a technique." He knew my knees were hurting me and came back over and told me there was no shame in taking a medical drop.

Meanwhile, we were gearing up and getting ready for a hasty attack, and I knew I had to now assault with the ruck as the RTO. I was crying by now. Not openly babbling, but I could not stop the tears from leaking into the corners of my eyes. It was at this moment that my Ranger buddy Rulla came over and quietly said, "Give me your ruck."

I said, "I am not quitting," and he said he wasn't going to let me, and he just wanted the ruck for the assault.

This simple and small act of brotherhood profoundly changed the entire way I saw soldiering. Rulla sacrificed when he didn't need to, for no other reason than to keep me in the fight. Those thirty minutes without the ruck on my back, and more importantly the love he showed, was exactly what I needed to turn the tide in my mind. At the end of the attack, the Ris told us we had a forced march of over ten miles through the night to get back to base. I grabbed my ruck and bit down on my lip. During that march I had a young RI, maybe twenty-four years old, march beside me and tell me for several miles that he didn't understand why the army let military police come here and waste a slot. He said I would never use this training. My anger at his comments numbed the pain, and the miles flew by. I am not sure today if he was just messing with me, but if I could reach him, I would tell him I have used what I learned in Ranger school every day of my life. Mountain phase taught me compassion and anger are combat multipliers.

As we reached Florida for the fourth and final phase, I remember a sign saying Ranger School Starts Here. I didn't get it that day, but it didn't take long for me to figure it out. As we left the mountain phase of Ranger school, I had been "promoted" to the class first sergeant, and upon arrival to Florida phase, I saw firsthand the impact of the student deaths that had occurred there just a month prior. Most of the cadre were in "acting" status, as most of the chain of command had been suspended pending a review of the accident. (Over a year later, a report would come out highlighting a cascade of bad-weather events that easily outmatched the lackluster risk mitigations and cavalier attitudes.)

We were told we could not get wet above the waist as a mitigation to avoid further deaths by hypothermia , and when this information spread, motivation skyrocketed, as we believed it was going to be an easy glide to graduation. However, our motivation was halted as we picked up over two truckloads of recycles from the previous class who looked like they were already dead. They explained that because we could not do all the rubber raft and deep swamp missions like normal, the RIs had cooked up continuous operational patrols. We would move to contact all day, then change leadership in the evening and move out to attack during the night. The next day we would do it again, and again, and again. Average sleep would be less than three hours a day.

At first, I was enjoying the swampy, warm weather of Florida in early April, as I was just happy to be on flat ground and giving my knees a slight breather. I even managed to get a "minor plus" rating while on point for spotting an enemy ambush before they were able to engage us. But something went wrong inside my intestines about halfway through, and I started cramping badly. I fought through the pain for about twenty-four hours not knowing if I had somehow injured myself or if my body was just failing on me. The next day my body let me know what was wrong.

I started defecating uncontrollably. I mean, I had zero ability to control my bowels. We were patrolling constantly, and I could not call time out. I soon had cut away my underwear and had feces in varying states caked to the insides of my legs down into my boots. I could not keep enough water inside my body and began to hallucinate. At one point while bounding away from incoming artillery,

I saw Jesus hovering over a tree. I kept running toward him until I had run past the point man and was reeled in by my squad.

My squad finally dimed me out to an RI when they thought I was going to die. When I told the RI I had the runs, he asked if I had to go now. I said I could shit on command, and he said, "Do it." He would not let me leave the patrol base, so right there, fifteen feet from the entire platoon chain of command planning a mission, I executed with true violence of action. The RI turned away and said, "Follow me, Ranger."

I was evacuated to the aid station with two others who had the same issue. The doctor said it was dysentery brought on by accidentally swallowing swamp water, or more likely the sharing of repacked dip (as rationed tobacco dip ran low, students would recycle tobacco and then even share recycled tobacco as times got harder) with hands that had not been washed in weeks.

Regardless of how I got it, I was miserable and scared they would try to put me out of Ranger school. I had not missed a second of school so far, and I was not about to fail out with just twenty-four hours left in patrolling. Luckily, the doctor gave us each several IVs and a few loperamide pills and told us we would rejoin the fight in the morning. We took full advantage of the medical clinic, showering that night and sleeping almost seven hours straight. The next morning, we felt brand new as we smeared some dirt on our faces and headed back to the platoon for the last mission. Today was going to be easy—not.

When we arrived back at the platoon, I could immediately tell they were near meltdown. They had been attacked and harassed all night and forced to break contact and move continually. They had not slept a wink, and guys were just passing out on their feet. I felt bad for whoever had to lead the last mission. That thought no sooner passed through my mind than I heard my name called out. I had drawn heavy weapons squad leader.

The platoon leader, platoon sergeant, and all of us four squad leaders were told we needed to pass this patrol because each of us was in danger of failing out. I couldn't believe it. I was the honor graduate from mountain phase, got a minor plus on point, and passed my patrol in Florida already. How could I be in danger of failing? Did they mix me up with someone? Was my night in the clinic taboo to graduation? I was in a panic. To make matters worse, my Ranger buddy Rulla, our normal point man and a Jedi of land navigation, was a walking zombie. Too tired to spit, he just let his Beech-Nut tobacco run down his chin, leaving his neck stained with flakes.

Any observer who might have happened by and watched our last mission would have rightly surmised that one school-crossing guard could have whupped all forty of us. We stumbled through the hot and muggy swamp like a gaggle of preschoolers and found our ambush site only by the grace of God. Getting my heavy weapons squad on line and concealed was like trying to pull the teeth of a tiger. When we finally got into position, the real struggle began.

As even the most casual reader will know, one of the keys to a good ambush is not moving and giving away your position. However, if you mix no movement while lying on the moist ground in the warm sun and no sleep for days, you can imagine what you will get. After an hour in the warm sun, troopers were falling asleep right and left. Each time required some leader to crawl over and wake the poor guy up. By the second hour, it was a lost cause. I untied the shoelace of my nearest M-60 gunner and told him, "When I pull your shoelace, you wake up and fire your entire belt." I only hoped it would give everyone enough time to wake up and attack.

Meanwhile, the platoon sergeant had crawled over and was lying beside me. He was struggling to stay awake, and I told him to rest his chin on the sharp edge of the weapon's carrying handle so the discomfort would help focus him. But this soon was failing, and he kept bouncing his head into the mud. I told him he needed this patrol to pass, to put his chin back on his weapon, and if he fell asleep again, I was going to punch his chin into the weapon. The next time he went out, I kept my promise and drilled him on the top of his head. His chin dug into the rear sight, and he woke up full of pain and anger. I thought he was going to hit me back, but he just said thanks.

As luck would have it, about thirty seconds later we both saw movement coming down the trail in front of us. No one had opened fire, and we realized the platoon leader might be out. The platoon sergeant looked at me and whispered, "Fire."

I yanked my gunner's shoelace, and like a charm, he woke up and dumped two hundred rounds from his M-60, jolting the entire line into action. The ambush happened. It didn't happen well, but the muscle memory of seventy-two days allowed the played-out warriors to dance one more time. Afterward we faced another forced march, but no one cared. Everyone knew this was the last march back to camp.

Ranger Class 5-95 Graduation – A weak 130lbs

The next day I found out only the platoon leader and platoon sergeant were close to failing. The squad leaders were stacked with the top graduates of the class. I was informed I would be the officer honor graduate, but even better, I was told that even though the platoon leader had failed, the platoon sergeant who had lain beside me got a "go" on the last patrol and would graduate. Florida phase taught me that when it is your turn to carry your battle buddies' weight, you do it like your life depends on it.

On graduation day from Ranger school, I didn't have any family there, but my ROTC instructor and college roommate made the journey. I stood there with only forty-two others from the original three hundred. I was barely 130 pounds and incapable of doing ten push-ups. But when they called out my name, gave me a Fairbairn-Sykes dagger, and pinned that black-and-gold tab on my sleeve, I was beaming and knew the army was going to be my home for a very long time. I was hooked on the army like crack cocaine.

CHAPTER 5:
MY HAITIAN VACATION (1995–1996)

*Great NCOs make good officers through
patience and support.*

AFTER RANGER SCHOOL, I took a week to gather my gear and put some weight back on and then headed off to Fort Polk, Louisiana. I had zero expectations of the state or the army post there. However, I was extremely excited to get to lead soldiers after almost a year in training. As I reached the small town of Leesville right outside the post, I remember thinking two things: there is not much around, and it sure is hot.

I signed into the 519th Military Police Battalion, which was about a five-hundred-person unit consisting of four companies, and was further assigned to the 258th MP company (one of the two large line companies). I was immediately assigned a platoon and jumped right into law enforcement and training. I should have been happy, but I wasn't. The platoon I had was a mere shell of a unit, with maybe eighteen soldiers, including only one NCO above the rank of E5 (sergeant). They had come back from a very tough mission in Panama during the Cuban Riots, where much of the unit had gotten wounded. They were good individuals, but most of them were months away from leaving the army, and there was no energy. I felt like my platoon sergeant (the lone senior NCO) and every other leader resisted all my efforts, from signing for equipment to aggressive training. Had I not left this platoon, I might not have gone very far in the army.

But fate and a Caribbean would-be dictator stepped in and helped me out. The initial US prevention of a coup in Haiti during 1994 was transitioning to the United Nations mission in Haiti. The 519th MP Battalion was tasked to send a company, the 204th Bladerunners, and they were short a platoon leader. I was asked if I would like to move over to this company and deploy as the third platoon leader. I jumped at the chance.

When I arrived at my new platoon, we had less than sixty days to train, certify, pack out, and deploy. It was chaos the first week as almost 30 percent of the company arrived at the same time I did. Back then, it was normal to have to shift personnel to bring a unit up to strength to deploy. A full military police company had about 170 soldiers in four line platoons and one headquarters platoon. Retirements, schools, injuries, moves, or folks just leaving the army quickly ate away combat power when a unit was looking to be gone for six months. Funny, back then we thought six months was a super long time to deploy.

It was during these sixty days of training and the six months in Haiti that I witnessed the incredible power of the noncommissioned officer. For the next quarter of a century and even to this day, I judge all NCOs I meet by the standard I was introduced to in third platoon. I remember how it started like it was yesterday. As I mentioned before, my previous platoon fought me on everything, sometimes nothing more than simply signing for the equipment. I was determined to not let my new platoon do the same to me no matter how difficult they made it.

So immediately after spending my first seven days counting every tent peg and screwdriver, as is the army standard, I meticulously accounted for over a thousand items in the platoon and created sub hand receipts (documents individuals signed acknowledging control and responsibility for the property). I made one for every NCO in the platoon and gave them to my new platoon sergeant. I told him I wanted these signed and on my desk the next day without any delay. SFC Kilpatrick, or SFC K, as he was called, simply looked up from his desk and said, "Yes sir." I left for the night feeling that tomorrow was going to be another fight with NCOs.

The next morning after physical training, I could not focus because I was constantly thinking about those hand receipts, and I was not seeing SFC K put out any guidance to the squad leaders about signing for the equipment. I bit my tongue all morning and into the afternoon. However, by 1500 (3 p.m.), I was so livid I could no longer contain myself. I shut the door to our office and engaged SFC K about why he had not put out any guidance, and was he trying to blow me off?

SFC K very calmly said, "Sir if you had looked in your in box this morning, you would have seen they were all signed last night."

I was speechless, embarrassed, and shocked. I did the only thing I could think to do. I walked out of the office, turned around, reopened the door, said "Good morning, Sergeant," and went straight to my box. I pulled them out and loudly thanked him for getting the receipts signed. He simply smiled at me broadly. I then sat down and told him about my troubles with my previous platoon's NCOs.

He responded, "Sir, there you did not have noncommissioned officers. Here you do. It is as simple as that." It was at this moment that my trust as a young lieutenant in my NCOs was galvanized.

The rest of our training went great, and soon we were on our way to Haiti. Back then units did just about everything themselves to deploy. We painted our own vehicles UN white, convoyed them hundreds of miles to the shipping port, and did the other thousand things to prepare. There were no contractors, garrison agencies, or services that we enjoyed years later during the war on terror. In many ways, these events brought our teams closer together. I got incredibly close to my squad leaders. My second squad leader, John Narcisse, would one day serve as my battalion CSM and become a lifelong friend. However, in Haiti it was my first squad leader, Mark Boucher, who taught me my second lifelong lesson about what it meant to be a good NCO, and his lesson came at a time when I was in real trouble.

The day started positively enough. I was taking three teams into the mountains surrounding the capital to recon several NAIs (named areas of interest) the outgoing unit had identified and that our company had not visited yet. We had only been in Haiti for a little over a week. One of the NAIs was a Catholic mission known to serve cheeseburgers and milkshakes, so I had little trouble finding volunteers to execute the mission. The patrol was beautiful as we rose above the hot and dirty streets of Port-au-Prince into the palm-covered ridges. We arrived at the mission just before lunch, but our hopes for a good lunch were dashed when we found out they were not serving lunch that day. So we decided to drive to a quiet ridge or outcropping with a view to eat our MREs.

We were at the extreme edge of our sector, and none of our maps covered the area. However, we could see a road off in the distance high above the capital that looked like a great place to eat and relax before making the movement back into the capital. So we headed off to try to find our way up. As fate would have it, the road we began to climb turned into a trail several miles up, then into a path barely wide enough for our vehicles. It was nearly vertical up on the passenger's side and nearly vertical down several hundred feet on the driver's side. I had just decided to find a place to turn around when the lead Humvee rounded a ninety-degree turn and the dirt on the outside of the trail gave way. The truck made it safely around the curve, but I now had one truck in front of the hazard and two behind it. I explored on foot farther up the trail and discovered it was a dead end.

After a few minutes of discussion between me and the team leaders, we decided to shore up the side of the road with some branches and dirt and then back the lead truck through the curve carefully. However, our plan was further complicated when we discovered the driver's side rear tire had blown out in the first pass, and it was impossible to change the tire on the trail. My driver felt confident he could make the turn, so everyone dismounted except him. I told him I didn't care if he tore the whole right side up, just stay away from the cliff side. His handling of the Humvee was perfect, but the flat tire acted like a monster scoop and dug into the shoring we had built with a vengeance. The truck lurched toward the ledge as the earth gave way, and before we could blink, it went over. Only the right-side CV joints digging into the earth as the truck sat at a

45 degree angle off the cliff, not my silly attempt to grab onto the truck, prevented it from rolling down the mountain.

As my driver scrambled up and out to safety, I took stock of the fine second lieutenant mess I had made. I had almost killed my driver and had my Humvee completely trapped beyond my assigned sector and outside of radio communications. I can remember thinking, *Well, this was a short and disappointing career in the army.* Even after hooking up a winch from the second vehicle and with nearly a dozen shovels and axes hard at work, we were no closer to a solution. I finally moved the third vehicle down the mountain a mile and got FM Comms with first squad, led by Staff Sergeant Mark Boucher.

Mark was one of the most seasoned NCOs in the unit, and I was relieved when he acknowledged my call to come meet me. Mark was one of the old breed. He could not spell "email" and typed like a chicken pecking for food. But if you needed to put a tent on top of a pine tree, he was the one who could figure a way to do it. He was a hard father to his soldiers and loved them fiercely. His humor would make a bouncer blush, and it often hid his true grit. But it was on this day that his never-quit attitude and loyalty taught me a lifelong lesson.

We linked up without issue, and I quickly explained the situation. We then maneuvered his squad up the trail to the site. He could tell I was scared that this might get me fired. He looked at the Humvee hanging off the edge and then turned to me and said, "Lieutenant, we will have you out in a jiffy."

I was immediately over my self-pity and began to refocus on solving the issue. He said exactly what I needed to hear. It was not until years later that he told me that after he said that, he turned to his squad and said, "We're fucked!" I still laugh when I think about how he made me feel better, his squad laugh, and focused the team all in less than ten seconds.

However, before we even began to formulate a new plan, a man rode up on a four-wheeler. Here we were, high up in the mountains where people didn't even have shoes, and a man had appeared on a four-wheeler. We didn't have a translator, but to our amazement, the man spoke perfect English. He said he had seen us at the Catholic mission earlier and appreciated all the help the army had provided in protecting them and assisting in their efforts. He then saw the Humvee stranded and asked if he could help. I said sure, but I was not sure there was much he could do because I was pretty sure neither a wrecker nor helicopter could reach us.

What happened next will sound like complete and utter fiction. But this man turned and yelled down the mountain, and within five minutes, we were surrounded by dozens of Haitians. Where they came from, I still do not know. They talked for a minute, and then off several of them went. Just a few minutes went by, and then rope, saws, axes, hammers, and another dozen men showed up. The Haitians began to hammer in anchors and tow ropes up the mountainside and cut trees and shored up the underside of the truck at the same time. Our troops dismounted, and for about an hour twenty American soldiers and several dozen Haitians became one team. The entire time, Staff Sergeant Boucher bridged the gap between peoples, led by example, and cared for and motivated every single person.

Within an hour we were ready to try moving the vehicle. It was almost completely anticlimactic. Teams pulled on the ropes and used large limbs as levers, and winches pulled in unison such that the truck popped up onto the trail on the first try. As soon as the Humvee settled, the whole group went crazy, as if we had just won the World Cup. It was an absolute celebration. We loaded down every truck with tens of villagers hanging all over, and we drove back down into the mountain village honking and singing like a mad wedding party. I think I hugged and kissed every Haitian in the town, and they were genuinely proud to have helped.

When we finally arrived back at our patrol base just before dark, my company commander said, "I heard you blew a tire. Glad you are back safe." He knew what had happened but also knew I had learned a lot that day and didn't need him to dwell on it. Captain Fawaz was a great commander.

Two of my most cherished Polaroid photos are that Humvee hanging off the road and the Haitian-laden party trucks of celebration afterward. For the rest of my career, I kept the first photo in my middle desk drawer to remind me, when I had to deal with something "dumb" a young officer did, that I was once him. Secondly, it reminded me of Mark Boucher, the old breed, who played hard, drank hard, fought hard, and loved life every minute. An army needs its old breeds to survive its well-meaning but naive lieutenants.

As I matured in the army, I realized the NCOs a young officer has greatly define him and his ability to lead throughout his career. I am proud to say I still am friends with over twenty members of third platoon. We've hunted together, cooked out together, and trained together, and they even stood as my honor guard when I married my wife. Great NCOs make and mold good officers.

CHAPTER 6:
RIDING THE CRAZY SPAIN (1996–1997)

⌒∾⌒

Success is not guaranteed, but great leaders give praise and take blame.

AFTER HAITI, I FELT like an old salt and walked around like I knew a thing or two. I had only been an officer for less than three years, but I was truly enjoying coming to work every day. I met amazing young officers, fantastic NCOs, and impressive soldiers. I just had not met any senior officers, majors and above, who inspired me. But to be honest, back in the 1990s you really didn't see many field-grade officers (major, lieutenant colonel, colonel) day to day as a second lieutenant. However, my few engagements with my battalion commander and majors left me totally unimpressed. This changed when Ted Spain took command.

Lieutenant Colonel Ted Spain was the exact opposite of the commander he replaced. He was fearless in both work and play. From the very beginning, it was clear he loved the army, loved soldiers, and was created by God to lead them. Lieutenant Colonel Spain spent every spare moment he could muster to be with troops. He would have breakfast with a private, run with a new NCO, or swing by a lieutenant's office to hear about their day. He never liked a lot of formality; he was focused on honest dialogue. As a young officer, I never feared his power; on the contrary, I was focused on not disappointing him. His trust in us, his dedication to our well-being, and his love of our profession drove me to give my absolute very best.

As I think back to this very impressionable time in my career, I can vividly remember countless events that demonstrated Lieutenant Colonel Spain's skill at making hard work fun. He would sometimes call a battalion-wide alert at 1500 hours on Friday with the alert message, "All hands report to the NCO Club." When you arrived, you would find hundreds of battalion members (not just officers and senior NCOs) laughing, joking, and truly bonding. When I arrived at the first one, there were three or four soldiers from the Second Armored Cavalry Regiment there having a beer. One of them was feeling froggy and yelled, "MPs suck."

The bar immediately went quiet. Lieutenant Colonel Spain turned and yelled back, "Kick their ass!"

In a split second, chairs hit the floor, and tables were flung to the other side of the room as a hundred MPs headed to crush these now-frozen-in-fear Second ACR soldiers.

At the last second, Lieutenant Colonel Spain jumped in front of the onslaught and stopped the charging herd. He then turned around to the Second ACR troopers and asked, "How do you like us now?"

The troopers made a bowing gesture to Lieutenant Colonel Spain and said, "We love MPs!"

Lieutenant Colonel Spain laughed, said, "Tequila for my friends," and hugged each one of them.

Instantly, those Second ACR troopers became part of the battalion and part of the party.

It was here that I learned the tradition of right arm night. It was when you invited your second-in-command, or right arm, to the club and bought him or her a drink. We did right arm nights monthly, and it was a time to bury bad feelings, discover new friends, and simply relax.

The author getting a pie in the face as part of one of Lieutenant Colonel Spain's fundraisers for the troops

It wasn't just at the club that Lieutenant Colonel Spain inspired his team. Though he was old for a battalion commander, he still did physical training with the unit. One time he showed up at physical training formation with buses and told everyone who wasn't scared to mount up. Of course, everyone piled on, and the buses drove ten miles out into the training area. Lieutenant Colonel Spain dismounted and said, "Follow me." When we went back to the edge of the main post after eight miles of running, he had us escorted back the last two miles by police cars and great fanfare. Years later, I found out that that run had really hurt him. But at the time, all I could think about as a twentysomething lieutenant was that if my battalion commander could do this, then I would die before I fell out.

Under his leadership, the battalion became a very tight unit that partied hard and trained even harder. This meant we were the perfect choice when a short-fuse mission came down to serve as the headquarters for four or five companies of National Guard executing a certification rotation in JRTC for their deployment to Bosnia. The Joint Readiness Training Center (JRTC) is a training venue at Fort Polk of hundreds of square miles sourced with role players, pretend enemy forces, and specially designed missions and challenges to both test and train units as part of deployment prepa-rations or yearly training goals. The swamps and heat of Louisiana coupled with the constant observations by scores of observer/con-trollers (part referee and part mentor) make the three-week "war" a hellish experience for most units and often invokes a visceral response just hearing the name Fort Polk and JRTC from most experienced soldiers.

When the call came down to the battalion to execute this short-notice mission, I had been on the battalion staff as an operations officer for about a month. After almost thirty months as a platoon leader, the army finally pried me away from the line, kicking and screaming, and placed me in a cubicle to both learn and help man-age the next level of our profession. Now, in the S3, or Operations Section, we coordinated for resources, managed tasks, and planned the month-to-month missions and objectives set forth by Lieutenant Colonel Spain. We kicked into overdrive to begin to plan and ex-ecute the rotation just three weeks away.

About one week before we were to head into the rotation, my boss came to me and told me I was also going to serve as the battalion's fire coordination officer. I did what any good young officer would do and said "Yes sir" and then went to find an NCO to tell me what I had just said yes to. The operations sergeant major, Sergeant Major Guyette, laughed and told me I was going to plan and coordinate artillery and aircraft support. I thought that sounded easy for a military police battalion because I had never used either of them in three years as a platoon leader.

I had just convinced myself that it would not be that hard when three artillery observer controllers (O/Cs) showed up at the battalion headquarters asking to see me. They immediately started hammering me with questions about quick-fire nets, preplanned fires, target reference points, range fans, and a million other things that made no sense to me. Once both the O/Cs and I got over the shock of how dumb I was at the task I had been given, the O/Cs said not to worry. They were going to teach me how to plan and execute fires while we were in the rotation. What they didn't say was how big a fire hose they were going to use to do it.

As we met the incoming National Guard companies who were assigned to our battalion and began to move into the rotation training area, it was obvious that this was going to be a hard mission. The companies were from various states, a mixed bag of quality and no support. As I tried to figure out what types of organic firepower we had within this ad hoc battalion, I discovered antitank weapons without ammunition, mortars unserviced, and dozens of unqualified gunners. The battalion was struggling to just get companies into a basic defense and establish some sort of command and control. Captain Chip Balk and I were developing plans, writing fragmentation orders (the way we issue guidance to subordinate units), fixing defensive gaps in both the physical line, and making up standard operating procedures. I would then sneak away and try to get some type of indirect fires established.

By the end of the second day, I had two 60mm mortars established and rehearsing base defense, and I was working to generate approved target lists for 155mm howitzers we had supporting our sector. These were incredibly large cannons capable of firing over ten miles and causing casualties up to a one-hundred-meter radius. True to their word, the fire support O/Cs rotated constantly in to see me, educate me, and then follow up with me to ensure I was progressing. They did not care that I had my day job in the plans section and base defense to help manage. They were completely focused on being able to kill the enemy with artillery. They were passionate in their craft, and it was easy to see why they were O/Cs. As the exercise progressed, it was clear that the rotation was geared toward peacekeeping, and my pace slowed to a daily grind of coffee, cranking out orders, walking the perimeter, and eating army chow. I figured the rotation would end without barely a shot fired. I forgot this was JRTC!

I was just shaving as the sun crept over the horizon when I saw a few folks running to staff and command tents looking for key leaders. I knew something was happening, so I quickly finished up my cold shave and headed into the command center. There I saw a series of reports streaming in from the companies reporting unusual movements of Serbian troops and roadblocks. A radio operator (RTO) yelled a report of a possible armored column moving down a major road toward our position. Our HHD commander, in charge of the base defense, could not raise the companies on the radio, and we had no visibility on any possible enemy. Lieutenant Colonel Spain grabbed me and said, "Schmick, go wake those SOBs and prepare to stop armor."

I ran down to each of the two entrances and alerted them to the possible enemy movement toward our position. It was like kicking a beehive as soldiers jumped into action readying what little they had to stop armor. I found two sections of Dragon gunners, shoulder-fired and wire-guided tank killers, only to discover they had deployed and spent the entire rotation so far without any rounds. Once company leadership arrived at the gates, I ran back to the command post to ready the 155mm howitzers in case we needed to fire in support of the base camp.

As I entered the command post, the intensity of the situation had increased several fold. Multiple squads from the companies on patrol were reporting possible small-arms fire at their locations, noncompliant civilian populations, and indications of pending military action. For context, at the time all belligerent parties had placed their heavy weapons in holding areas and were not allowed to remove them, under the NATO agreement for peacekeeping. However, a single report arrived that changed the entire character of the situation.

A single squad on patrol near a Serbian heavy weapons holding area reported the Serbians were loading out all their tanks and armored personnel carriers. The patrol further reported taking fire from them and was pulling back to a safer overwatch. All eyes turned to the battalion commander, Lieutenant Colonel Spain, as this was a clear, unprovoked attack on US forces. The staff lawyer gave his assessment of the use of force rules, and the intelligence officer stated he had no external information or intelligence confirming or denying hostile intent. This was before the digital age; the only information the commander had was radio traffic from a young sergeant and the knowledge that he could not stop an armored force once it loaded up and moved. He asked and received verification from the unit on the ground that the Serbians were continuing to prep for deployment out of the holding area.

I had already plotted the holding area days previously as a possible target and sent a warning order five minutes earlier to the firing unit of a possible mission thanks to great training from the artillery NCOs previously. I stood quietly in the corner of the command post watching my battalion commander do the combat calculus in his head. Lieutenant Colonel Spain did it quickly and realized he only had one method to stop the attack. He walked over to me and said, "Schmick, can you hit that holding area?"

I said I had twelve tubes of 155mm ready to fire if directed. I saw out of the corner of my eye one of the artillery O/Cs, and he was like a proud father of a boy getting ready to hit a home run.

Lieutenant Colonel Spain locked eyes with me and said, "Hit 'em."

"Fire mission, fire mission, fire mission. Twelve rounds HE, TRP 007, troops and armor stationery in the open. Fire." I had waited for weeks to say those words. I was shaking with excitement as the tubes reported the shot, the expected detonation, and the confirmation of impact from the squad on-site.

However, the squad on-site for some reason could not assess the effectiveness of the strike. They stated they could not give us a BDA (battle damage assessment) and were unsure of the damage on the enemy tanks. But the artillery seemed to have hit them. There was a brief discussion between Lieutenant Colonel Spain and his operations officer, Major Tim Weathersbee, about whether they could risk the tanks departing the holding area. Lieutenant Colonel Spain said we had already hit them, and then he turned to me and told me to hit them again.

"Repeat. Repeat fire mission over," I echoed over the radio.

The firing process occurred again, and my artillery O/C, a young E-6, was absolutely thrilled to finally watch his pet project get to put steel on target. The squad finally called in with a BDA assessment of all targets destroyed.

I high-fived Captain Chip Balk, and Major Weathersbee came over and said, "Great job, Glenn."

I had just put massive hurt on the bad guys in the way of twenty-four 155mm high-explosive rounds on an area the size of a football field. I didn't think about how much death and destruction I had just unleashed; I simply celebrated the fact that I had been given a new task, learned it, and when needed used it to stop the enemy cold. The rest of the enemy attacks seemed to melt away and not materialize much. I naturally assumed our rapid response with overwhelming force had broken their spirit.

A few days after my victorious hour of glory, the entire battalion leadership and staff, along with the company leadership, gathered for the after-action review (AAR) done by our JRTC O/C teams. This process was done for every unit in the rotation to capture positive and negative issues through the operation from an objective viewpoint for units to know what to continue to train on when they left. This concept of the AAR has saved thousands of lives over the years and is a valuable part of our professional culture. The value of the AAR on me that day cannot be overstated.

For most of the first hour of the AAR, the O/C feedback was not very flattering, to say the least. They centered on the companies going through the validation exercise, as they were the primary recipients of the training. I was admiring how dirty my body and uniform were after a few weeks without a shower when I heard the familiar voice of one of my artillery O/Cs start to talk. He started off by acknowledging the ad hoc nature of our task force and the lack of qualified artillery personnel. He then began to talk about the event just two days prior. He stated how we had effectively planned, and when required, had rapidly placed effective fires on a target with devastating effects. The graders at the site assessed over a dozen armored vehicles destroyed and scores of dead Serbian soldiers. Everyone on the staff was looking over at me, and I was as proud as a peacock.

Then a senior officer O/C stepped forward and said, "Now for the bad news." None of those Serbian forces were attempting to leave the holding area, attack coalition forces, or do anything other than routine authorized training. We had just committed a violation of the rules of engagement and a possible war crime. In an instant my pride evaporated as my mind struggled to understand what he was saying. He had to have the wrong event, as they had fired on US troops and were loading out to attack. I was stunned. The evaluator went on to explain that the squad patrolling nearby thought they heard a gunshot and then witnessed Serbians loading out their vehicles. However, there had been no shot, and the Serbians were conducting routine inventory and maintenance on their stored equipment per the peace accord. Furthermore, the squad called for fire based on their assumption and without establishing and maintaining eyes on the objective. In other words, they thought they were shot at, took off down the road, called for fire, and never went back until the rounds hit.

Lieutenant Colonel Spain cut in and said he had made the call and accepted 100 percent responsibility for the incident. As I took everything in, I realized my perception and the battalion's perception of the battle was completely incorrect and based on inaccurate information.

I didn't hear anything that was said after that. The AAR went on for another hour, but I had drifted inward questioning what I had done and how wrong I had been. Worse yet, I remembered how thrilled I was when I did it.

I was so focused on myself that I didn't notice Lieutenant Colonel Spain until he was standing in front of me. Somehow the AAR had ended, and most of the folks were already headed back outside. I jumped up and blurted out, "I am sorry, sir. I messed up."

Lieutenant Colonel Spain looked down at me and said, "That just isn't true. Schmick, I made the call, and when I did, I needed you to be able to make it happen, and you did." He told me the call he made was wrong in hindsight, but if given the same information again, he would make it again. He was upset that the squad had failed, but he also said they were trying to do the right thing. They were just not trained or disciplined enough to do it. Here was a senior leader talking to a lieutenant like I was his son. He accepted responsibility, put the incident in perspective, and built me back up all in the span of one minute.

The entire two weeks in the rotation taught me two lifelong lessons in the span of just a few minutes. First, in combat you can do everything right from your perspective and still be dead wrong. And second, great leaders take responsibility and care for their subordinates no matter what.

CHAPTER 7

HOUSEHOLD SIX AND WAR DOG SIX (1998)

❧

Leadership is influencing others positively whether
you can order them or not

ABOUT THE TIME MY platoon leader time was ending and my time on staff began, I met the love of my life. To say I knew she was the one is a complete understatement. After almost three years of limited dating, I was at a point where I was dating several women and convinced I would just enjoy my time before I departed Fort Polk. However, by sheer coincidence I ended up going out to eat with another officer I had not seen in weeks, and Angela was our server at a local restaurant. Unknown to me at the time, she was working between college classes and had just broken up with her fiancée. I didn't discover any of this, because I was terrified. She was the most beautiful woman I had ever seen, and when she spoke in her sweet southern accent, my mind went blank. As we left after dinner, I told my friend I was completely a mess and should have asked for her number.

My buddy laughed and said he had to hit the latrine before we left. When I met him at the car, he handed me her number. He had recently gotten engaged and said he wanted to play matchmaker. As for Angela, she never gave her number out and certainly never dated soldiers. But on this night, something made her give her number to my friend for me. Maybe it was her recent breakup, or maybe it was the pathetic way I went blank when she smiled. Either way, my buddy became my hero that night. I soon called her, and we went out on our first date.

On our second date, I told her I was going to marry her. Angela went home and told her mother that I was a little weird and she might not go out with me again. As for me, I knew she was the one, and I ended every other relationship.

At the time of this writing, that was twenty-four years ago, and she has been my only girl since that chance meeting. Little did we know at the time that a year later we would be army newlyweds and in command of a company while still a lieutenant.

By the beginning of Lieutenant Colonel Spain's second year in command, I had asked officially for Angela's hand. Until then, I had never really thought about how families fit into the army. I could say all the buzzwords, but I didn't understand the challenges of balancing both. Lieutenant Colonel Spain had a secret weapon that I never recognized until after Angela and I were engaged. R'ami Spain, his amazing wife, was just as active and engaged in the unit as any officer. When she heard Angela and I were engaged, she immediately reached out to her. She was not intimidating, demanding, or any of those Hollywood stereotypes you hear about. R'ami took time to explain how the army worked in the eyes of a spouse, how to utilize systems and care for one another. Angela later told me that R'ami played a large part in giving her the confidence and assurance that becoming an army spouse was a great adventure but totally doable.

We were married on a beautiful June day, just one year after we met, in the main post chapel with members of my old platoon as my honor guard in a full military wedding. Angela looked amazing as I took her arm, and she whispered with a smile, "If you mess this up, I will kill you." She of course was joking, I think, but she had witnessed my army buddies kidnap me from the rehearsal dinner and throw me in the trunk of an old, unmarked police cruiser. And she had already been told that morning about how hungover I was from a night I still cannot totally remember beyond someone buying me a full glass of German whiskey and questioning whether I was really a Ranger. Now, as I stood there in the ninety-degree heat of a chapel with a broken air conditioner, Angela made it clear I had better be a Ranger and gut it out.

Angela didn't kill me, and we began our new life together at Fort Polk setting up a new house with the expectation that we would soon leave so I could attend the Advance Course (a six-month course for officers as they make captain).

However, all that changed one night during a no-notice alert to send the 204th MP Company to Desert Thunder. Captain Niave Vernon (Now Brigadier General Niave Knell) was preparing to rapidly deploy her company to Kuwait as Iraq was demonstrating hostile intent. My boss, Major Weathersbee, told me the battalion commander needed to see me and ask a favor the next morning. I was so excited. I thought a platoon leader must not be able to deploy, so they wanted me to take the platoon. I was so hyped, I could not sleep.

However, when I went into Lieutenant Colonel Spain's office the next day, what he asked came out of left field. He asked if I could take over Headquarters and Headquarters Detachment. I was shocked and disappointed all at once. First, HHD, as it was called, was the least desirable unit to command because it was the unit that had all the staff and none of the line troops. Second, I had just stood in the change of command ceremony the previous week for a new HHD commander, and I was expecting to go off to Advanced Course within a few months. Why would Lieutenant Colonel Spain ask me to stay another year, not go to school, and then command an HHD? What had I done to be so derailed from a successful path?

When Lieutenant Colonel Spain finished telling me this, he asked if I had any questions. I asked the only thing I thought of, which was, when would I take over? He responded that he needed me to take over in twenty minutes.

If I was confused when I walked in, I was now hopelessly lost. I sheepishly asked why. I knew the current commander had only been in command for three days.

Lieutenant Colonel Spain said the commander had taken over the company on Friday, dismissed the unit, and then drove to Houston. Then on his way back from Houston he was pulled over by the state patrol and had five pounds of marijuana and a loaded handgun on him. As a military police officer with almost seventeen years in, there was no bigger betrayal.

I now understood. The battalion had been dealt a serious blow to both our credibility, honor, and ability to function. To recover quickly, Lieutenant Colonel Spain was taking a huge chance by putting a lieutenant in command. In an instant I knew I had to take that command. Lieutenant Colonel Spain knew I was concerned about not being able to command a line unit. He told me I would serve as the HHD commander for eleven months. This was a critical move because the standard to be considered key and developmentally (KD'd) complete as a captain was twelve months of command. By allowing me to come out of command before twelve months, he was ensuring I would get a chance to command again.

Angela was as excited and terrified as I was about taking over a company. Just four months earlier, we were both single, and now we were not only married but in charge of a seventy-soldier unit and responsible for their families and the welfare of the troops. I told Angela over the phone the news and then took command in a conference room with no fanfare. Afterward, I walked down to the company with my first sergeant (it was his first day too), and we stood outside until someone arrived to let us in because we didn't even have keys. The next days and weeks were a blur as I set up a cot in my office and tried to figure out so many things. I had never been exposed much to budget, staff calls, family readiness, unit status reports, commander's actions, etc.

Taking command of HHD was exactly what I needed at this time in my career. I didn't realize it at the time, but I might have been a good trooper and I may have been tactically sound, but I did not know much about commanding or professional relationships. My wife and 519th Viper teammates would use this opportunity to teach me both.

One of the unique aspects of an HHD is that most of your troops work for another officer who does not report to you. In fact, inside the HHD are the battalion commander, command sergeant major, and nearly a dozen officers. The primary mission of the HHD commander is to ensure the staff and commander can function effectively to command the rest of the battalion. For me this meant my natural tendency to use direct leadership and focus on planning and execution of missions was not the tool that I needed for this job. At the same time, this environment created time for me to focus more on soldiers and their families beyond just mission execution.

Angela "bloods" (pounds the author's rank) Glenn
during his promotion to captain

Angela and I got to know our soldiers and their families on a level I had not as a single platoon leader. Angela would write birthday cards, make baby gifts, or just ask them about their birthdays or anniversaries. I was amazed at how much their attitudes would change and how much they would look forward to her visits to the office. I began to try to keep up with her and learn about their families as well. Now, don't get me wrong. I knew my platoon exceptionally well, but it was not the same. As a platoon leader, I got to know my troops. As a commander, my wife was teaching me to know their families, goals, and dreams beyond being a soldier. It was powerful.

At the same time, I was struggling to learn to lead without giving orders. Day to day, I only had five people whom I could direct. Since I was a lieutenant, everyone else worked daily for someone else who outranked me. Whether they worked in the S1 (personnel section), S2 (Intelligence section), or one of the other four sections, they all had an officer who was focused on completing priorities set by Lieutenant Colonel Spain, the field-grade officers, or the command sergeant major. My requirements were a distant third or fourth on the list at best. This meant I had to grow and become a better communicator, influencer, and teammate just to succeed.

I had tremendous captain mentors who did not see me as competition, but instead as a battle buddy and little brother. They put their arm around me and taught me effective peer mentorship. The two line company commanders who spent so much time with me, Niave Knell and Jeff Stewart, went on to be a general officer and colonel respectively. I knew they would be successful then because they just had the gift of leadership and a joy and a sense of humor in everything they did. However, sometimes the best lessons come from bad moments. It was Major Tim Weatherbee who taught me one of the most profound lessons as a young commander.

Tim was what we called a quiet professional. He was the operations officer, and he had been my boss when I worked in the S3 shop. Now that I was a commander, Major Weathersbee was in my unit. He was one of two field-grade officers within the battalion beyond

Lieutenant Colonel Spain. He had taught me how to write better and coached me through my first development of a PowerPoint presentation as a young staff officer. I remember asking why I needed to make a PowerPoint presentation, and why couldn't the battalion commander just read the hundred pages of detailed work I had done to develop an operation? I still laugh when I think back to his simple answer. The boss does not have time to read your hundred pages of detail; you get three slides to convince him that you do not need to do the hundred pages over again.

As powerful as those lessons were, the lesson he taught me via email and the professionalism of going face-to-face still resound today. It happened one Friday afternoon, just after I took an update on some minor maintenance tasks that each section needed to complete. I don't remember what it was, only that several of the sections failed to accomplish the task even though I provided specific guidance and allotted several weeks to execute a one- or two-hour project. The battalion executive officer, the other field grade, charged with "running" the staff, was completely useless. Any time I went to see him, he would put his hand up in my face and say, "Before you speak, does what you are going to say involve work for me? If so, the answer is no."

I had tried to be transparent and predictive and to provide maximum time for the staff to complete their company requirements, understanding that their main fight was the battalion effort. But on this day I concluded that I was not being respected and that sections believed they could blow me off and not do the tasks I asked. I was mad, and I was determined not to let this continue. Email at the time (1997) was still brand new, and only primary leaders and staff had accounts. I decided the best way to set a standard and put my foot down was to address all the officers and NCOs at once via email. So I fired one off.

I was still fired up about the email when my phone rang just minutes after I hit Send. Major Weatherbee asked me to come to his office now. I grabbed my hat and headed the two hundred yards to the headquarters, excited knowing he must have seen my email and was going to help drive change in the staff. He and his wonderful wife, Susan, had always been mentors to Angela and me since we joined the staff, and I was sure he had a solution. So I was stunned when I entered his office and he didn't ask me to sit down but instead began to very quietly chew my ass.

Major Weatherbee told me he was disappointed that I had decided to chew him out in front of every other leader in the battalion. He stated he had been nothing but supportive of all my efforts, and I had no right to challenge him publicly. He further told me it was unprofessional to not come to him face-to-face to discuss the issues. He closed by reminding me that not one time had I raised this issue with him. I stood there biting my lip waiting for him to finish so I could correct his misinterpretation. It was obvious he had not understood my email, and I needed to educate him.

When he finished dressing me down, I explained to him he had misread my email. I told him about the executive officer not supporting me and several of the other sections not completing what I asked even after giving them maximum time. I told him he had always been a tremendous role model, and the email was not directed at him. It was simply him misreading what I wrote.

He decided I was not getting it. He stopped me and said, "Everything you just said is not what you wrote. I want you to read this email out loud right now."

I stepped around his desk and read it as he directed.

As I read it out loud, I realized three profound lessons about email and effective communication overall. These lessons were so pivotal that I later would teach them to colonels as an instructor at war college. First, all the context that I had in my head when I was writing the email never made it into the email itself. I assumed the reader would have the same perspective I did at the time of writing. Second, I thought it was the reader's job to understand what I was saying, as opposed to the sender's (my) job to effectively communicate. And finally, I realized I could not account for how the reader would process what I wrote if I was not there to further discuss it. He had taught me the dangers of lazy communication.

As soon as I finished reading the email, I realized how wrong I had been. I apologized profusely and I saw his hard gaze soften as he switched to a mentorship role again. I had offended someone I truly respected, and he turned it into a learning moment—simply incredible. I learned to never send an angry email. Write it, place it in your draft file, and then read it the next day. As I did this, I found I deleted 99 percent of them. Second, I learned an email sent is not a task complete. Instead, it simply makes the sender feel good. In the world of army high stress, nothing built teams and cleared confusion better than meeting face-to-face. And if you could not meet face-to-face, the email you sent had to be clear, concise, and stand alone.

I went into my HHD command believing it was a diversion from becoming a successful officer. Instead, I discovered holes in my ability to lead and manage larger formations. My wife taught me the importance of families within the army. When I looked at her, I knew I would do anything for her, and how my organization treated her mattered. Therefore, I had to focus on soldiers' families as well to achieve the same sense of pride, satisfaction, and security we felt in the army. My peers, especially Major Weathersbee, taught me that only basic leadership was giving orders and being accountable. The next level of leadership was influencing others positively whether you could order them or not. This meant being a good citizen, teammate, and follower. As Angela and I departed Fort Polk in the fall of 1998 heading to Officer Advanced Course, I realized I could not have asked for a better assignment than the 519th Vipers.

CHAPTER 8:

SUPERSTARS EQUAL SUPER LONG HOURS (1999–2001)

Sometimes to be a good leader, you need to lead less and follow more.

OUR TIME AT OFFICER Advance Course flew by. I became incredibly close to many of my classmates, and to this day we still visit one another and share in life's journey. Angela and I were sad to leave Fort McClellan, Alabama, but super excited and nervous to be heading off to the Ninety-Fifth Military Police Battalion (nicknamed the Superstars) in Germany. All we knew was we were going to be assigned to Stuttgart, Germany, where I would command an MP company that was several hours away from the battalion headquarters. I thought it would be a great adventure and that it would be so much easier to command a second time. I was wrong.

For those who have never been to Germany—or for that matter, served in the US Army in Germany—it is hard to describe how different it was from any other experience to this point in our career. It was before the internet, smartphones, the euro, or GPS. Germany was a totally different culture, and the US Army looked, operated, and felt different from the army I had left in the States. Getting gas, buying music, or just traveling around were skills that had to be taught by soldiers and family members to new soldiers and family members. It was a demanding place for young families and this young officer.

I had been assigned to command the 554th Military Police Company, which had the mission to be prepared for war while providing law enforcement and security for the Stuttgart military community and United States European Command Headquarters. For those who are not familiar with the two sides of the military police competencies, the skills required to excel in the tactical side of military police and the skills required to excel in the law enforcement role are very different. Since the inception of the branch, there has always been a tension between training for one at the expense of the other.

The 554th MP Company's challenge was exasperated by three additional aspects that made this command brutal. The first two aspects were products of time and distance. In Europe no military police companies were colocated with other MP units. So, this meant that each MP unit had to provide law enforcement for its community twenty-four hours a day for 365 days of the year. At any given time, regardless of holidays, training, or major events, there was a steady, never-ending requirement to provide qualified, armed, and ready police. This may sound simple, but it is not. Our commitment was about thirty MPs spread over three shifts, which meant you needed to have forty-five working to allow each soldier to have two days off a week. Additionally, you had to rotate all these workers to different shifts every few weeks to ensure safety, equity, and training. This began to be impossible to achieve.

Secondly, 554th was unique compared to other MP companies in Europe because it was assigned four platoons' worth of equipment but only authorized three platoons' worth of soldiers. I never discovered why, but I believe it was because the fourth platoon had been split off to the European Command Headquarters there in Stuttgart to serve as a protective services detachment (bodyguards). And of course, no unit was ever filled with all its required soldiers, meaning on any given day, soldiers had to maintain, service, and keep track of twice as much equipment as any other MP unit. However, it was the third aspect of this command that had the greatest impact on both the troops and me. The battalion commander was a brilliant, driven, decisive leader who almost killed us all.

Lieutenant Colonel Dave Lemauk was the closest leader I had ever seen to a machine. He was a small man in stature at under five foot eight and weighing less than 150 pounds. However, he easily ran two miles in under twelve minutes and could run at a sub-eight-minute mile pace with a thirty-five-pound pack on his back effortlessly for miles on end. It was clear that the Ranger Regiment scroll on his right shoulder and the President's 100 patch were earned through his tremendous physical achievements. He rarely slept and employed two full-time soldiers as drivers to sustain his eighteen-hour-a-day schedule. Though married, he had no children or outside distractions to impact his ability and desire to focus on the battalion continuously.

Lieutenant Colonel Lemauk's physical fitness and endurance were matched only by his mental capabilities. On top of racing sport bikes for fun, it was rumored he spoke five languages. But it was his depth of army knowledge and his ability to recall paragraphs and chapters from literally hundreds of books, regulations, and orders that made him unmatched in his capacity to consume and retain information. He could routinely recall unit statistics from two or three weeks before and compare them to current numbers while in a meeting with company commanders. He was by every measure exactly what you would want to command a demanding unit. The challenge in being superhuman is being able to get the best out of those who are not.

As I took command of the WarDawgs, I discovered a unit in complete crisis. The unit had more leaders under adverse investigation or on physical/mental profile (*profile* is a medical slang term in the army for a medical limitation emplaced on a soldier due to an injury or illness) than fit and functioning leaders. I could not muster more than two dozen NCOs or officers within the entire company. The outgoing commander was under investigation, the unit had no purpose or vision, and morale was the worst I had ever seen. There were great soldiers and leaders in the unit, I would discover, but the environment they faced daily was a crushing tidal wave of tasks, poor synchronization, and a complete lack of training. My time in command would be a challenge between my desire to succeed professionally and my desire to care for troops. This would define my army future.

In the post–Cold War era of 1999, the army in Europe was struggling. Resources were drawing down from a perceived victorious theater, while the appetite of allies and army senior leaders to execute operations had not reduced. To say the force was stretched would be a gross understatement. For months on end, I had to work my MPs on twelve-hour shifts (which once you factor in physical training, drawing weapons, and finishing paperwork was actually sixteen hours). To make matters worse, soldiers were only getting one day off a week and leaders only one day off every two weeks. To maintain weapons qualifications, we would load soldiers into vans after they worked all night and drive them four hours to crew-served weapons range. They would sleep on the way there and then on the way back so they could immediately draw weapons and patrol the next night for law enforcement. This meant they were on duty for thirty-six hours straight.

I continually failed to convince my battalion commander of how desperate circumstances were within the unit. He simply set a standard and expected it to be met no matter the changes in the environment or resources. He never wavered once. As an example, if a soldier had an accident in a government vehicle, he required the entire unit to go to two people in every vehicle. For MPs, this meant our mission requirement doubled. He would make units hold this standard until the police investigation was complete, the unit did an accident board, and then the entire chain of command drove up to the battalion headquarters and briefed the results and recommendations to avoid future issues. It did not matter if the soldier hit a pole with a mirror or a curb with a tire; the result was two-person patrols.

At one point, we were so short of personnel, I had to put cooks and mechanics riding along with MPs just to not violate his orders.

I already had every officer and myself and the first sergeant riding along as well. I slept in my office for days on end and even developed a staph infection due to it. I was working my folks so hard, reasonable accidents were occurring, and the battalion's answer was to work them even harder. I finally drove up to the battalion headquarters and asked the battalion commander to approve my risk assessment (a document leaders were required to use to manage the risk of missions and operations). I felt the risk to soldiers and to the mission was so great that it was beyond my level to approve it. The army guidance stated it was up to the next level of command to review the risks when they reached this level of danger. I felt this was a way to try to provide relief without breaking trust with my commander. However, Lieutenant Colonel Lemauk told me he would not read or sign the risk matrix. He stated my purpose was to mitigate risk and accomplish the missions I was given. If I could not do that, he would find someone who could.

I left his headquarters defeated and confused. How could we risk soldiers without thinking about why? Why was I not able to shape his understanding? I started to realize my commander's strength was his weakness. Everything was important to him, and everything was a no-fail mission. He would call my office at 0515 to discuss a soldier's evaluation and then call me at the house at 2030 (8:30 p.m.) the same day and ask me why I had gone home so early. Weapons qualification, vehicle maintenance, award processing, physical fitness—every single topic and task that was assigned to our unit had to be perfect. There were six other company commanders serving with me at the time, and five of them left the army shortly after their commands. The pace was maddening.

I would have failed within a year if it were not for a completely committed wife and some tremendous leaders within the company. My wife and I rarely spent more than an hour a day checking up, but Angela still gave up her weekends to come with me as we visited newly arrived soldiers and their families. Some were super nice, and some spouses took out years of frustration on us on the first day we met. But we promised we would meet every new family, and we kept that promise. My frustration with my battalion and the pace of Europe was tempered every time I spent time with my soldiers. Whether it was my motivating lieutenants who never lost their positivity or my rock-solid Operations Section and platoon sergeants, they always fired me up.

My two junior platoon leaders, Jody Lupo and Janette Kautzman, were disciplined, smart and incredibly caring for soldiers. They loved taking care of their platoons and shouldered more than their weight in every task. It is no wonder both of them went on to become lieutenant colonels and highly successful. They were initially led by my executive officer, Jeff Toczylowski, who was a fearless, fun-loving leader who absolutely could find the good in any situation. The only time I ever saw him embarrassed or sad was when, at a party at my house, he spilled an entire glass of red wine on my wife's new white couch and carpet. Anyone else would have been banned from our house, but Angela just laughed because Jeff was like her brother. Jeff was my right-hand man until he went to Selection and joined Special Forces in my second year of command. Years later we got to meet Jeff's family and share with them how much he meant to us.

My great officers were matched by a very tight group of superb NCOs. My operations sergeant, Kevin Schneider, was a former first sergeant, and like John Narcisse, he helped cement in my mind how an NCO was supposed to lead. He was an informal leader within the company and was both my mentor and friend. Whether it was his counseling, maintenance, tactical knowledge, or ability to care for troops, Kevin did everything as the army intended it. He led a group of tremendous E6s and E7s who bore the brunt of the workload to try to keep the troops from going crazy and the missions from failing. In the absolute hardest times, my NCOs just kept going and tried to make it fun for everyone.

MSG Kevin Schneider's last day working before retirement in March 2001

Sergeant First Class Rex Sprague and his wife were a perfect example of the NCOs we had within our company. He had been an infantryman and a medic in his younger years but now served as a military police platoon sergeant. His salt-and-pepper hair hid his young spirit and drive to care for soldiers. When he arrived at the unit, Angela and I went to meet them at their house one weekend. Rex and his wife had prepared lunch and would not hear of us not sitting down for a bite to eat. He taught me that genuinely caring for others was a combat multiplier. Rex never lost that spirit, even after he retired a few years later and deployed to Iraq as a contractor. He was mortally wounded in an ambush and still drove his vehicle out of the kill zone, saving others. He died as he lived, putting others before himself.

Even with great officers and NCOs like Rex within the company, the demands of my battalion coupled with my own personal drive to make the best unit were crushing soldiers. I was beginning to realize, very slowly, that I was developing like my battalion commander. Even though I didn't like it, I was not willing to accept failure, quit, or reduce my expectations and standards. I was at a crossroads in my career even if I didn't recognize it. And it was at this crossroads that, for the third time, a major taught me wisdom with just a few words.

Major Jennifer Gray was our battalion operations officer. Day to day she handled operations of the largest MP battalion in the free world. Each week when I made the two-hour drive to the battalion for Super Tuesday, which was a marathon of four to six meetings with the battalion commander, a visit to her office was mandatory. I went there to try to understand what was going on above me, shape what I could, and gain appreciation for the challenges above me. It was during one of these visits that I pushed my unit's agenda hard, and Major Gray upended my perspective with just one sentence.

The battalion was preparing for a range density week. During this week, three times a year, the entire battalion would send individuals and teams who needed to qualify on crew weapons to a series of ranges run by various units at one location four to six hours away from their home station. It was common to have hundreds of soldiers on each range simultaneously firing M-60 machine guns, MK-19 grenade launchers, M2 .50-caliber machine guns, M203 grenade launchers, and M249 machine guns along with grenade and claymore mine ranges. All this occurred while these same units continued to execute law enforcement back at their home stations. In order to accomplish both tasks, every unit had to do excruciating planning and flawless execution.

As I entered Major Gray's office, I had just received several short-notice taskings from her office requiring me to change multiple firing orders and convoys going to the range density. The effects on my unit were unavoidable. I had used this training event as a center of focus for my tired unit for almost three months. We had moved simulators into the barracks and even cut the top off a Humvee and put it in the attic so soldiers could train with crew-served weapons inside. I had targeted this range week for us to become number one in the battalion in weapons qualification. It would be a huge motivational win for the unit, not to mention a feather in my cap. But this new wave of taskings would undo all we had worked to achieve.

I walked into Major Gray's office, and she looked tired. But as was her nature, she stopped what she was doing and asked how she could help a company commander. I told her about how hard I had been training my team for this density and how focused they were on it. I told her the taskings I had just got would totally disrupt all my detailed planning and, worse yet, would destroy my chances at winning top company in weapons qualification.

As soon as I finished that sentence, Major Gray cut me off and said, "Well, it is a good thing Lieutenant Colonel Lemauk commands seven companies and not just one."

I looked at her, lost, and said, "Ma'am?"

She told me the battalion needed to be ready to utilize seven "ok" companies and not one great company and six broken ones. "If I have to break your perfect plans to keep every company in the fight, I will do it every time."

Like a lightning bolt, it hit me. I asked to sit down and said, "Ma'am, you are dropping serious knowledge on me."

One of the key tenets of being a good leader in the army was being a good follower. It was my job to make my next-higher unit successful even if it meant my unit was less successful. My problems were not the problems my battalion needed to focus on, but my battalion's problems were what I needed to focus my unit on. Once I digested that, I realized I had to let go of some of my goals for my company to support my battalion. I could that see my battalion commander struggled with doing this at his level, and if I did not change, I would be just like him. His spirit was in the right place, but his drive to achieve his goals and his boss's goals were crushing to everyone around him. I had to change, and it was at the conclusion of this second command that I started to formulate what I would later call the long game versus the short game.

CHAPTER 9:
WAR FROM HOME
(2001–2002)

❧

*As a leader in crisis, most of what you will do you were
never trained for—fall back on Problem-Solving 101.*

AS WE FINISHED COMMAND in Stuttgart, I hoped to go to be an instructor at the MP school. However, Europe was short MP officers, and they decided they would move Angela and me for one year to Kaiserslautern, where I would be a force protection officer at a mainly civilian transportation element. It seemed like it was going to be a sleepy little assignment, but after my near exhaustion in company command, I was ready for a little relaxation. We arrived in July, and in August I was sent to a course on seaport security in Amsterdam, where I met another MP officer, Victor Baez-An. We both were newly minted force protection officers and wondering how we would not die of boredom in our new jobs. Just two weeks later, everything would change for us and the rest of the world.

I was down in my new office in the basement of the First Transportation Movement Control Agency beginning to close for the evening when my door burst open and an operations lieutenant said, "Have you heard anything?"

I said, "Heard anything about what?"

He looked at me in shock and then blurted out that we were under attack, and I should come to the command center to see the news. I almost thought he was joking, but the look in his eyes told me this was not a prank. I ran upstairs and entered the command center, where dozens of people were staring at the TV. The World Trade Center towers had just collapsed.

The deputy commander, a lieutenant colonel, directed a meeting in thirty minutes in which I was to tell the team what we were supposed to do. Our commander was trapped in another country as all air flights were grounded, and we had thousands of shipments all over Europe moving at this very moment. I ran down to my office and grabbed the force protection books I had been left by the previous officer and realized they were useless. They were writing for the Cold War and talked about evacuating families from Europe as the Russians attacked. We were in virgin territory. I told the team I had no answers, but there would be changes coming rapidly for sure.

The changes I thought were coming in the next few weeks came in the next few hours as overnight the American military attempted to protect itself and its families. We locked gates, took side mirrors off trucks to make vehicle inspection tools, set up fire pits to warm guards, and even used our own vehicles to teach young soldiers how to inspect vehicles for bombs. Traffic backed up outside every post, and basic supplies like food and mail could not get through the security rings. We were not prepared, either technically or mentally, for the changes 9/11 ushered in. Angela and I lived off post, and it was scary to think we were now outside America, and someone wanted Americans dead. However, the Germans were incredible hosts. Thousands of German civilians swarmed to tell us they were on our side, and the German police and military rose to protect every US installation. It was a special time of special trust between our countries.

No sooner did the initial shock wear off than the tasks began to flow in. A week before 9/11, I could not have been a more "unimportant" staff officer than the force protection chief. After 9/11 I found myself involved in almost every aspect of daily operations. In just a matter of weeks, we had mitigated most of the big risks to our forces in Europe, and we began to look toward offensive operations and shifting forces toward the Middle East. It was in this context that I got my first real taste of the power of the US Army in times of crisis and the power of the US dollar.

It was near the end of the week, and my boss came in and told me I needed to pack a suit and be prepared to fly to Romania the following week. He told me there was a port we were looking at using to begin transporting equipment and supplies to the Middle East theater of operation, and we needed to conduct a reconnaissance to see if it would work. I was going to lead the force protection and security portion of the reconnaissance. My first thought was, *Finally, I am getting to do something proactive to help the US strike back.* My second thought was that I needed to buy a suit!

At this time, Romania was not a NATO member, and few army personnel hard spent any time there. When we arrived in the capital, we were treated like VIPs, with diplomatic escorts at the airport and an official greeting from their Ministry of Defense and Transportation. As a young captain who had only ever gone anywhere in a Humvee and sleeping bag, this was super cool. Looking back, I failed to grasp any of the operational or strategic implications of what we were going to do. I simply was not capable of thinking beyond the tactical environment.

Of course, the tactical environment around us at the capital was enough to keep my mind racing. After our meetings with the official ministries, which took place in very old government buildings, we scheduled a meeting with the businessmen who oversaw key aspects of the transportation. These men arrived in Mercedes SUV caravans and were surrounded by heavily armed security. I first thought these men had to be senior presidential officials. However, one of the other members of the recon team leaned over and said, "Mafia."

While we were whickering our way through old Soviet-style bureaucracy, capitalist mobsters, and the technical details of transit, I was still taken aback by the strange contrasts and beauty of this country. Gypsies still roamed the streets in wagons, large areas of the countryside had no trees thanks to Soviet strip mining, and the nascent democratic economy was struggling to gain traction. However, there was a genuine love and admiration for the US and a desire to help us avenge 9/11.

One unique aspect of Romania during this time was the dogs. More specifically, stray dogs. They were everywhere, and most of them looked like they were barely alive. When we checked into our hotel on the coast later that week, I got up before the sun to take a run like I always did. As I got ready to leave the hotel entrance, the young woman behind the desk frantically tried to communicate something to me. However, she spoke no English, and I spoke no Romanian. She pointed to one of the dogs outside the hotel and made a biting sign with her hand. I smiled at her, showed her I had my knife on my side, and told her I wasn't scared of dogs. Then off I went.

I didn't get more than a few hundred meters before I noticed I had picked up a tail. It was not a human tail, but a posse of dogs was following me. Every minute I ran, my traveling party seemed to grow rapidly. By the time I reached a mile, there were well over thirty dogs following me. However, they were not jogging with me. They were acting like they were going to run me down. Each time I would yell "Back!" or lunge at one, they would recede less and less. My pride told me to keep running, but when one dog got too close and I drew my knife to scare him, he did not run away. He held his ground and growled. I knew this was my signal to retreat. I ran the distance back to the hotel in record time, as now forty-plus dogs took turns pushing my pace.

When I finally leaped back through the hotel entrance, the same young front-desk girl pushed a newspaper under my nose as I stood trying to catch my breath. She pointed to a specific article, but I didn't understand. Another guest was heading to breakfast, and he leaned over and said, "She is pointing to an article about a man who was attacked and killed by stray dogs last night on his way home from a bar."

My naive and arrogant smile was gone. I was tired, humbled, and thankful for this young lady's concern for me. As exciting as our trip had been up to this moment, it paled in comparison to the adventure that awaited us at the port of Constanta.

As our recon delegation, about eight strong including transportation, medical, finance, and legal specialists, arrived at the port, I could not help feeling like I was in a Cold War movie. We were officially greeted by the press snapping photos and official speeches, and then led into a conference room. Inside the room was a massive table with tiny Romanian and US flags everywhere and our names on place settings. A staff of six beautifully tall young women dressed in too-small business skirts and six-inch-high heels served us drinks. Meanwhile, the air was thick with smoke from almost a dozen Romanians chain-smoking nonstop. As one of the most junior members of the team, I sat way down the side of the table and for the most part was able to just take in this incredible scene.

After well over an hour of discussions at the front of the table, the meeting started getting into the specifics of what we needed to make this port viable. When it was my turn, I rattled off a substantial list of requirements fresh from the port security course I had attended in Amsterdam just a few months prior. Most of the items, like secure housing locations, controlled access, and local national vetting, were all met with positive nods and smiles. When I started discussing the need for waterborne protection and sea exclusion zones, the Romanians had a feverish sidebar discussion. Surprisingly, they then asked if we could go to the navy yard and see if they had what we needed. Before I could respond, the senior army delegate said absolutely and told me and the finance representative to go.

In a matter of minutes, I was standing in front of a row of more than ten warships docked neatly in their births. My Romanian guide was the commander of the equivalent of their SEAL team and looked like he was fifty years old but hard enough to kill me with a toothpick. He asked me what we were looking for. My mind flashed back to the movie *Spies Like Us* when Dan Aykroyd and Chevy Chase shop for weapons. I stated that first we needed to defend against surface attacks like the one that hit the USS *Cole*. When I said that, my Romanian SEAL yelled something to the nearest ship, and a whistle blew. In an instant men ran and manned their machine guns and cannons. I was getting the full show of their capability. In the back of my mind, I kept wondering, *Does this guy know I am an army military police who doesn't know port from starboard?*

In a matter of five minutes, I had picked two one-hundred-meter-long ships along with two fast attack rubber assault boats. The Romanian officer stated that these craft would be very expensive to operate and could cost as much as $8,000 a day. I hadn't thought about the cost until that moment. I looked back to the finance agent with me and asked if we could afford it. He looked at me like I was two and said, "You are good; I can write a check up to two million dollars today."

Wow, I thought. *I am seeing an entire side of the army I never knew existed.*

Little did I know at the time that this trip was just one of many I would take in the coming months. Before I finished my tour in Europe, we had set the conditions for the global war on terror and the next two decades of my career.

CHAPTER 10:

THIRD TIME IS NOT A CHARM (2002–2005)

❧

There are bad leaders, good leaders who make bad decisions, and thousands who have to live with them both

AS MY THREE-YEAR TOUR ended in Germany in the spring of 2002, there was no hint of the major operations to come in Afghanistan and Iraq. Consequently, I focused on trying to go to Fort Leonard Wood as my next assignment to teach the Captains Career Course, which I thought I was perfect for after two commands. However, there were no slots available, so Angela and I decided to try to get closer to her parents in Louisiana. My branch manager said he could get me to Monroe, Louisiana, if I wanted to be a recruiting company commander. Of course, I had heard horror stories about recruiting, but I figured it was like anywhere else in the army where you could define your success. So we headed off to take a third company completely unaware that it would be the most brutal assignment I had ever experienced.

I honestly could write an entire book of unbelievable stories about my time in recruiting. Whether it was my first sergeant getting relieved on my first day in command, sending seventeen applicants to join the army in one day and the process only allowing one to make it, or even the silly requirement to scan used urine test kit results on flatbed scanners before we sent applicants to join. I could write about young college graduates with teaching degrees who could not score the minimum mental aptitude or the wonderful young and dedicated who wanted to join so much but could not because of a childhood ailment. Instead, I will focus on the sheer stress of the environment matched only in my career by my worst days in Iraq.

First, it must be said that recruiting for the military has always been hard. However, by the time Angela and I settled into our house in West Monroe, Louisiana, in the summer of 2002, US service members were beginning to die weekly as our rapid special operations strikes into Afghanistan were growing into something larger. Americans were patriotic toward "their" military, but few were looking to place their loved ones in harm's way. The army was struggling to make the required recruitment numbers to sustain the all-volunteer force. This was a paramount concern for us, because the army was and is so good because no one serving was drafted or forced to serve. The army was willing to do whatever it took to find the recruits it needed.

I felt supremely confident, having commanded two companies already, that this command would not be that hard. I have never been more wrong in my assessment of anything. First, my company was responsible for seventeen thousand square miles with recruiting stations in five separate towns in two states. By regulation, my first sergeant or I had to visit each station every week. This meant on average I stayed in hotels at least a day or two almost forty weeks out of each year. I ran my operations on a laptop behind the wheel of a car as I drove three to four hours a day.

At the same time, the socioeconomic environment of northeastern Louisiana and northwestern Mississippi made it hard to find qualified young men and women. Our data showed us that we had to meet several hundred people just to find one to join. Low education levels, high casual drug use, and minor criminal infractions eliminated over 75 percent of the people we met. Those who were left were the same ones heavily recruited by colleges and successful businesses. The idea of turning down one of these opportunities to go to war was not a popular decision.

To add to the challenge, most of my recruiters were physically and mentally exhausted beyond belief. I had about twenty-eight recruiters from the rank of E5 to E7. On a normal week, if they were successful, they worked twelve hours a day during the week and six hours on Saturday. Of course, this did not include transporting applicants to hotels or dealing with constant emergencies. I only promised them two Sundays off a month. This meant at best my recruiters worked sixty-six hours a week. But if you were to contact them today and ask how many hours they worked, seventy-five to eighty was normal.

For most of them, recruiting was just a one-time assignment for three years away from their normal army career endured to increase their chances of promotion, so the hardworking conditions were seen as payment. This historically brutal pace was exasperated by the army's decision shortly after I arrived to extend recruiter assignments from three to four years. The army was not getting the recruits it needed and thought keeping experienced recruiters on longer was a solution. This meant overnight all my recruiters were unceremoniously told they would execute another year at this breakneck pace.

If a hard operational environment and extremely long hours were not enough, individual recruiters faced constant stress as success or failure was measured monthly, and positive leadership techniques failed to survive first contact. Recruiting command was simply a toxic place to work. In my thirty-six months, I had the top company in my battalion all three years. However, I only made my assigned mission three months. This meant threats, abuse, and punishment rained down constantly. Driving a business bottom-line sales focus was contrary to army leadership values but not to the army's recruiting command. From the top down it was a tremendously abusive environment that placed incredible pressure below without any empathy. This meant many good leaders succumbed to bad practices to attempt to make the mission.

As an example, one of my battalion commanders who was a great infantry officer decided that if a recruiter had an applicant who did not ship off to basic training, the recruiter would come to see him for a letter of reprimand. This meant if the applicant got an unpaid parking ticket, a sick mom, or a broken leg that kept him from going to basic, the recruiter had to come. It was a totally misguided attempt to drive change within the organization. This may not sound like a big deal until you realize the recruiter had to come on Saturday in his dress uniform at 9 a.m. with his commander or first sergeant to the battalion headquarters in New Orleans, which was four hours away. This meant almost every other Saturday for nine months, I got up at 4 a.m., drove four hours to New Orleans, stood beside a recruiter while they spent five minutes with the battalion commander, and then drove four hours back. The running joke was, "Beatings will continue until morale improves."

As I struggled daily to find some balance between leadership strategies, the chaos of recruiting, respect, empathy, and making the mission during a time of war, I found I was losing the battle. Angela, now pregnant, rarely ever saw me, and when we were together, I was on the phone, moody, or stressed out. She was alone in a strange town without the army spouse support she had known when living with other military families. The impacted disks in my back were becoming debilitating as the constant driving, hotels, poor eating, and stress led to several failed medical procedures, which had me barely able to stand. I could not stomach the thought of failing, but the drive to succeed at the mission was costing me my family, my health, and the leader I thought I was.

It was during this environment that I met my first bad brigade command sergeant major. Up until this stage of my career, I had met hard CSMs and plenty who were hard to work with as an officer. But they were always clearly acting out of their dedication to their role as senior advisers and the keepers of standards. No matter how much we disagreed about a specific topic, it was always clear we shared reverence for the army and its success. However, when I met the Fifth Brigade CSM, I knew he was not cut from the same cloth.

I first met him just three days after I took command. On the day I took command, the battalion commander relieved my first sergeant and then told me I had to fly down to San Antonio in two days to brief at a Bottom 10 briefing. When I asked what that was, he told me that the bottom ten companies out of forty-five had to go brief the BDE commander and CSM. Then he added as an afterthought that I was number forty-five. Since I had not been to any training before taking command, I struggled to learn as much as possible about basic recruiting, my unit, and my unit's failures in just forty-eight hours.

During the Bottom 10 brief, I sat and watched nine company commanders and first sergeants get destroyed in public for an hour each. The brigade commander would open the attack but then allow his CSM to completely dominate the dialogue and bludgeon the units. It only took me an hour to realize there was no right answer to any question and this was simply designed to abuse and shame commanders into wanting to not come back. Of course, taking an ass chewing is part of the army, but this was something more. This was personal. I watched as the company commander before me had his leadership book thrown at him by the brigade CSM. Here was a prior service captain with nineteen years of service including Desert Storm getting a book thrown at him in public. To his credit he showed more maturity than the brigade leadership and held his professionalism. When it was my turn, the CSM asked how long I had been in command, and when I told him three days, he simply said, "Don't come back." There was no welcome, no asking about me as an individual, and no sign of leadership. I made a note of how to never welcome a teammate.

Several months later this same brigade command team came to our battalion headquarters to take our quarterly production brief. This quarterly brief was the forum used to review units' success or failure. However, the brigade command team used it to go to the worst units and inflict their style of motivation and abuse. I was not too concerned because I had put in over 120 percent of my required aggregate mission, moving the company out of the bottom ten in just ninety days, but had not met the minimum requirements in each category to technically make the mission as defined. But after spending almost every waking minute trying to recruit and retain recruited candidates, I was confident I was doing everything I could. My wife had gone to her parents for an extended visit, and I was basically living in my office. I was strung out along with every other company commander present, but it was clear everyone was committed to giving it their best. Until their visit...

I learned the night before that I was going to go first in the briefing order, which I thought was odd because in my mind I had done well and achieved 120 percent of my missioned enlistments. It was obvious just seconds into my brief that the brigade sergeant major thought otherwise. This sergeant major was the same one who just three months ago I watched throw books at combat veterans and belittle them in public when I first took command. Unlike our first meeting, where I was just a spectator, he was now engaging me directly. His technique was simple. Within recruiting there were metrics for everything from the number of appointments made in an hour to the market share of enlistments per zip code. He would just find a metric, any metric that you were not good at, and attack. Unlike other aggressive leaders I met in the army up to then, he made his attacks personal.

I stood at the podium jabbing and parrying statistics and fielding questions for well over an hour from the commander and sergeant major. Even though there were dozens of other staff and commanders in the room, the sergeant major spoke with complete disregard for professionalism or fear of accountability. I attempted to focus my efforts on the objective facts and not his antics. However, as we neared the completion of the second hour, the constant verbal attacks and sarcasm had worn away my defenses. I was physically and emotionally drained before I arrived, and by now I was demoralized as well. What I failed to realize until it was too late was that I was also mad as hell.

The moment I lost control is as crystal clear today as it was over twenty years ago. The brigade sergeant major asked why my soldiers were simply not able to make the exact mission they were given after they had been given all the tools required for success. I took a deep breath and prepared to cover the same ground I had been briefing for over two hours when he cut me off and said, "Obviously you embrace and teach failure—no need to respond." He then looked to the brigade commander and began to recommend the next commander step up to brief. But he never finished his sentence because he had tripped a trigger that I was unable or unwilling to stop.

In an instant, I left the podium and crossed ten yards to the head table. I stopped right in front of the sergeant major and yelled, "Sergeant, you asked a question, and you are going to get an answer." Calling a command sergeant major a sergeant was totally disrespectful, but it was worse because I was waving an eighteen-inch piece of wood in his face at the same time. What I did not realize until just then was that I was so mad at his comment that I snapped the bottom edge of the podium off and was now wielding it like a weapon. However, I was determined to have my voice heard. I shouted, "I was working over one hundred hours a week. My NCOs were determined. We were dedicated and had no quit in us. However, this constant beatdown is neither effective as motivation nor warranted."

At the end of my outburst, I stood there frozen in place with a chunk of broken wood in front of me. I was greeted with complete silence.

After what seemed to be a minute but was just a few seconds, the brigade commander said, "Next briefer" without even raising his voice.

That was it. Not another word was spoken. I went back to my seat and sat silently for the next four hours of the briefing. I expected to get fired for sure after my total lack of emotional control and professionalism. At the end of the event, the brigade command team left, and my battalion commander told me to come into his office. He asked if I was OK, and I apologized for letting the CSM get the best of me. He smiled and said the brigade commander just wanted to make sure I was "good to go." I responded, "Yes sir. I am good to go," and I departed.

The day the Schmicks gave up command after three long years

I served in this recruiting pressure cooker for another thirty-three months until I pinned major. It did not break me, but it came close. The effects of this command were like a cross of Ranger school exhaustion and the depression and fear of future deployments combined. I learned there are a few bad leaders in the army, there are good leaders who make bad decisions, and there are thousands of amazing leaders in the army who just bite down on the bit and grin and bear it to get the mission done. Recruiting left scars on me but made me ask myself what type of leader I wanted to be and the cost I was willing to pay. Little did I know the cost was going to go up.

CHAPTER 11:

SAND, HURRICANES, AND DEATH (2005-2006)

A strong family and good friends are critical to fight depression and loneliness.

AS MY THIRD YEAR in recruiting ended in 2005, I was finally given orders to Iraq. I had lobbied hard, afraid the war would end before I got there. It may sound odd, but I explained it once as spending your entire career in the NFL—you would want to go to the Super Bowl and test your mettle. I was a baby major itching to get into the fight. All the normal army schooling was postponed to support manpower forward in the war, so I was allowed to deploy without attending Intermediate Leaders Education (a year-long master's level education for new majors). Angela and Hunter moved back to be close to her family at Fort Polk. I received orders to Third Army HQ in Kuwait with the understanding that I would deploy forward from there. The next fifteen months would profoundly change the way I saw the world.

When I arrived in Kuwait, my desire to get to Iraq was immediately smashed when I was told I was going to remain in Kuwait on the Third Army staff. I was told I was going to be the chief of US Customs for the military—a fancy title for ensuring equipment and personnel redeploying did not violate US Customs and Border Control laws by smuggling illegal items or bringing back unclean and unsafe items. Since I wanted to get into the fight, this was the last thing I wanted to do. But no matter how much I tried to find a way out of it, I was told to suck it up. This led to almost eleven months of grinding administrative and bureaucratic tasks sprinkled with only a few trips across the berm into Iraq.

Camp Arifjan in Kuwait was the epitome of every major Hollywood movie showing the big headquarters far away from the fight, almost in an alternate reality. We had silverware at our chow hall, a large Olympic pool, a movie theater, and a shopping complex. It was the major hub for contractors, equipment, and command and control for everything supporting the war in Iraq. This sprawling complex had tens of thousands of people moving in, around, and through daily. Its purpose was clear and essential to the war effort, but I couldn't help but feel ashamed as I walked around without body armor, worked out at the air-conditioned gym, and made PowerPoint slides about issues that even I didn't find valuable. The Third Army was General Patton's unit in World War II, and I could only imagine him rolling over in his grave as he saw his colors flying over a headquarters not in the fight during war.

The leadership of the Third Army Provost Marshal Shop, where I worked, was marginal at best and toxic at worst. By 2005 we were beginning to see officers who would have never been promoted gain rank due to the need to fill tasks. The colonels and lieutenant colonels in our shop were nepotistic, and most functioned from Atlanta, Georgia, just emailing forward their work for us in Kuwait to do. Some of them even worked day on and day off there because they claimed not to have enough computers. Our senior colonel would fly in the last day of the month from Atlanta, shop for a day, have a few meetings, and then fly out in the first two days of the next month. He would then get two months of combat pay for seventy-two hours in Kuwait. All the heavy lifting was done by two other junior majors and me forward. Dave Koonce and Chris Heberer, whom I had known both before, became my battle buddies and confidants. I would have gone crazy there if it were not for their maturity, humor, and friendship.

In my time in Kuwait, I learned two lessons that stuck with me for the rest of my career that made the poor leadership worth it. First, I was forced to learn effective communication. This concept is hard to explain, but it involves spending more time evaluating all aspects around a question to effectively answer the question. If someone asks, "Is the weather going to be bad?" you could not just say yes, it is going to rain. You had to analyze why this person was asking, bad for what purpose, what was the duration of the time period they were focusing on, etc. During my time on the Third Army staff, we spent dozens of late nights trying to take hundreds of pages of information and condense them into a single page for a general officer or congressional delegation. Each time, the three of us would spend hours reviewing the meeting notes where the requirement came from, who the general was who had asked the question, and from what lens they were interested in the issue. This skill, as painful as it was, would become a strength for me as I became a senior leader later in my career. The second lesson I learned was how very different the war was across the berm.

Autor's first trip into Iraq in 2005

In my role as the senior army customs agent, it was my job along with Central Command's (CENTCOM) agent to certify customs programs including those in Iraq. So on a few occasions, I would fly around Iraq and inspect programs. Until then the most excitement I had experienced was on a helicopter that had to make an emergency landing in the desert at the border. But on one trip I paired up with a marine lieutenant colonel out of CENTCOM to inspect the marine units in country and travel into Al Anbar province in western Iraq. It would not be my last time going there, but it was the most stunning.

We flew into Al Taqaddum (TQ) airbase, a marine strongpoint, under the cover of night. As we arrived, the first thing I noticed was the lack of lights. Everything was dark, unlike the large forward operating bases in Baghdad. We stumbled through the dark to find our marine security force point of contact and began to make our way to the VIP quarters, which turned out to be a tent with nothing but cots. However, before we got there, the base took incoming artillery, and we were ushered off to the nearest bunker. As I stood in this bunker in the dark, I was still not processing my surroundings very well. Looking back, I think I was just not accepting what my senses were experiencing.

As I struggled internally to gain situational awareness and switch my mind to a combat footing, I looked around the bunker. We were near the Marine Security Forces barracks, so we were surrounded by six or eight very young marines. One marine stood there in his brown boxers with body armor and fuzzy rabbit slippers on his feet. Next to him was a female marine, no older than nineteen, leaning against the wall looking very bored as a few loud explosions echoed in the distance. The group bitched and complained about the inconvenience of being attacked and picked on one another's state of dress. I stood there in silence. I was in their world, and I was clearly just a tourist. Here again I witnessed the adrenaline, camaraderie, and love that would come to mark a generation.

The state of Al Taqaddum was in direct contrast to the neat, clean, and orderly Arifjan in Kuwait. Piles of ammunition were lying around on the ground, port-a-johns were overflowing, and trash blew by in waves. It was clear the marines here had little time for creature comforts. Over the years I have seen many good and not-so-good marine units. Absolutely no different than the army. However, I tend to judge marines to those TQ marines of 2005. They were resolute, focused, and committed to the fight no matter how hard it got. After spending a few days there I was laser focused on getting to Iraq. However, first I had to get out of TQ.

Our last major stop during this mission was Al Asad Airbase, farther east in Al Anbar. In 2005, the roads were getting hammered with EIDs, so flying was the only option for us. All aircraft flew out at night to avoid ground fire, so we were scheduled to leave around midnight. It was so eerie to walk out onto the flight line there at night. There were no white lights at all, and the ground was simply lit with chemlights to mark specific locations. One color for where to walk, one color for runway, one color for casualty collection points, etc. At the same time, the sound was deafening. There were dozens of aircraft in various states of landing and departure. You had no choice but to surrender your fears about accidents and just submit yourself to following the chemlights and hoping for the best.

We walked for a short distance and arrived about forty yards behind the little glowing light that marked the inside of a CH-46 helicopter. The CH-46 was the marine two-propeller workhorse utility helicopter, which had been part of the corps for almost forty years. These birds were getting pushed to their endurance limit, and it was not lost on me that I was younger than the aircraft I was boarding. However, I was mainly thinking how crazy it was that I was boarding a helicopter with only a pistol, traveling a route I didn't know, at night with no communication. My safety was completely up to someone else. I decided right there that from then on, I would always carry a long rifle.

I came to this conclusion just as I entered the dark cabin of the 46 and was greeted by the smell of oil and a slippery floor. I looked down and could just make out some type of fluid leaking across the metal. I tapped the crew chief and pointed to the fluid. As soon as he saw it, he turned and started pushing me and everyone off, yelling to get away. Of course, he really didn't have to say it twice, because we bolted off the bird like sprinters and didn't stop until we were back to the green chemlight forty yards behind the bird. Once there I turned around and saw him hitting something violently inside the craft with a hammer or wrench.

I was glad we escaped with our lives and started wondering how we would find another ride when that same crew chief stepped back out and waved us forward again. I was thinking, *Hell no, I am not going back there.* But he continued to wave, so we reluctantly reboarded the aircraft. I leaned over to him in the dark and asked him what had just happened. He smiled and said, "These old birds sprung a lot of leaks. This one was just hydraulic fluid, not fuel."

If I was not nervous about flying that night before, I sure was now. I said a prayer, gritted my teeth, and made my peace as the old bird ascended into the dark.

I finished that flight and the rest of my business in Iraq without incident and returned to Kuwait only to find you don't have to be shot at to find yourself in pain. Over several months, I found no fulfillment in being caught in the daily staff grind while the war and the rest of the world marched on. If this was not enough, Hurricane Katrina hit Louisiana, where Angela and Hunter were living. Here I was in the desert trying to order fans, batteries, and water to get to my family who were without power. How could I not be there to help and protect my family? I felt guilty knowing that what I was doing daily was not worth the separation from my family in need. But Angela would not entertain my worries. As would become her hallmark, she simply said, "I have my big-girl panties on, and we will be just fine." She never wanted me to worry about anything at home no matter how serious it was when I was deployed.

Shortly after the effects of Katrina subsided, I was hit with a second gut punch at chow one night in early November. I was eating alone and grabbed a *Stars and Stripes* newspaper, which were free at the entrance, to read while I was eating. The newspaper kept you up on the daily war, sports, and a few events back home. Additionally, you felt like you were connected to generations of previous soldiers who had read this newspaper in countless wars before. I had just started to read the paper when I caught a picture of Jeff Toczylowski, my old XO from the 554th MP Company who went to Special Forces. I was wondering what crazy stunt he had done to get into the *Stars and Stripes*. Before I finished that thought, my eyes locked onto the headline. It said, "Deceased Special Forces Soldier Leaves $100,000 and a Letter to His Family and Friends." My mind could not comprehend what I saw. Even as I feverously read the short article, none of it made any sense to me. Jeff was Captain America, and there was no way he could even get harmed, let alone killed. I remember getting up and walking out of the chow hall. Then I just walked out into the desert and cried.

Jeff was everything you pray your child will become, and I could not fathom him being taken from us. That night I called Angela, and we just cried on the phone together. I only discovered his death because his last letter had made the news, unlike the now-dozens killed each week in the war. In typical fashion, Jeff wanted his passing to be a party. He wrote that he died doing what he loved and wanted all his family, friends, and military friends to party together and celebrate his life, so he left $100,000 to make it happen. A year after he passed, near Veteran's Day, he had his party in Las Vegas, Nevada. Angela and I were there, just a few weeks after I returned from my fifteen-month deployment. We had decided to go and then not go a dozen times. It was so emotional, so surreal, and even today that weekend is a blur in my consciousness. But it was amazing in so many ways to meet his team and his family and to share in more than a few crazy stories. However, at the time of his death, I could not fathom the party. Only my sadness.

I pushed harder to find a way to get to Iraq and shake the feeling of not doing my part. I now had lost two amazing soldiers from my previous company as Rex Sprague, one of my best platoon sergeants, was killed driving his truck out of a kill zone as well. Rex and his wife were the nicest military couple we ever encountered during command. I called everyone including my old battalion, the 519th, to see when they would be deploying. As luck would have it, 519th was in Iraq and needed an operations officer in the coming months. After a lot of working the angles and explaining to my wife, I was able to work the system to allow me to change units and duty location without going back to the States. The only catch was I would do over eleven months in Kuwait and then go straight to Iraq for an additional three-plus months. I didn't care because I needed to find purpose, I needed to go to Iraq, and I needed to be relevant in the fight.

However, just ninety days before I was to go to Iraq, I got the Red Cross notification that nobody wants to hear. My father, who suffered from demons in the form of longtime alcohol and drug abuse, had died alone in Texas. For a second time, I found myself in the desert crying. My tears barely made a mark on the burning grains. I felt guilty again. Had I done enough to help him? Had I told him I loved him? I pushed aside all my emotions and made my list of tasks to get done.

I had to fly home the next day and then drive to San Antonio. Once I arrived, I realized no one else could or would be the executor of his estate, so I had to do all those tasks in just ten days and then return to Kuwait and get ready to deploy north. I didn't allow myself a chance to grieve. I simply switched to mission mode. I paid his debts, settled his accounts, divided his possessions, spent a few precious days with Hunter and Angela, and then departed back to the desert. I didn't realize it then, but I was already out of balance.

Kuwait had been a brutal assignment in many ways. I had seen waste, I had seen misery, I had seen my family threatened, and I had seen both family and dear friends die. In many ways my faith was tested, and my service in the army was questioned. But my closest battle buddies and my incredible wife got me through the sand, hurricanes, and death. As I boarded the C-141 for the flight up to Baghdad, I prayed I had enough left in the tank for what was to come.

CHAPTER 12:
DANTE'S INFERNO (2006)

❦

Nothing can prepare your mind for the realities of combat operations—just accept it.

AS I LANDED AT BIAP (Baghdad International Airport) late in the afternoon on 15 July, I was filled with nervous energy and anticipation. I dragged my bags outside the military passenger terminal and looked around for a unit representative. I had coordinated with the battalion the previous week, and they knew I was coming in on this bird. I waited for an hour, and no one showed up. I called the battalion operations office in a very unhappy mood and asked why I didn't have a ride. The operations NCO, Sergeant First Class Boulis, very apologetically said they were working to get a patrol to me now. I was mad and told him this was unsatisfactory since they knew I was coming on this flight for almost three days. He simply apologized again.

After another forty minutes, a patrol finally showed up which was led by the 258th MP company commander, Captain Shevin Denmark, and First Sergeant Koch. I thought they were really trying to make up for leaving me here for two hours if they sent a company command team to get me. I told Shevin it was fucked up that they could not figure out how to get me on time. He apologized and very quietly said the patrol that was tasked to get me got hit on the way with an IED, and the squad leader, Sergeant Andres J. Contreras, was killed. His soft words punched me in the face. Here I was showing my ass by complaining about not having a red-carpet reception when an amazing young soldier had lost his life simply trying to come pick me up. There is almost not a week that goes by that I do not reflect on Sergeant Contreras's sacrifice and the lesson he taught me even though we never met: "Be humble in service."

As I got into the up-armored Humvee, still in shock from the news, I realized I could not even figure out how to lock the doors. These vehicles were completely foreign to me. I still remember the young private showing me how to open and close the door. I bet he was thinking, "Oh, great. Our new operations officer cannot even open a door." As we traveled through Baghdad on the now-infamous Route Irish and countless other roads against traffic and over medians, all the while forcing traffic away from us, it all felt surreal to me. I went from a place where you expect flawless timing and convenience to a place where you were trying to not die in a period of just hours. I thought about the old saying, "Be careful what you wish for."

My introduction to my new unit and Camp Rustamiyah was like drinking the ocean through a straw in six seconds. Rustamiyah was a dirty and crowded camp on the southeast side of Baghdad just a stone's throw away from both the Tigris River and Sadr City. Multiple army units, contractors, and Iraqi formations were crammed into a base that was surrounded by heavy traffic, burning waste, and trash. The first aspect that struck me was the color tan. Everything from sand, buildings, sandbags, uniforms, and vehicles was tan. There simply were no other colors present.

The first thing I did when I arrived at the FOB (forward operating base) was to find the operations NCO I was a jerk to on the phone and apologize. He was incredibly professional in his response and said, "I am sorry I could not tell you over the phone, sir. It was an unsecure line."

The 519th was just as "crammed" as Rustamiyah. A typical MP battalion was designed to manage three to five companies, or roughly 500 to 800 hundred soldiers. The 519th had anywhere from seven to ten companies at a time depending on rotations, swelling ranks to 1,200 to 1,400 troops. These troops were spread out across two division battle spaces and several hundred miles. My S3, Operations Section, had over twelve officers and NCOs just operating the tactical operations center. The battalion was responsible for training and mentoring the Iraqi police in all of east Baghdad and as far south as Al Diwaniyah. In addition, both of the two most violent districts, Sadr City and Adamiyah, belonged to the battalion's mission set.

I barely slept the first night but got up early and headed to the outdoor range on the FOB to qualify with a rifle. I was not issued one in Kuwait, so getting armed was my first priority. The same operations NCO I met yesterday, SFC Boulis, took me out to the small twenty-five-meter target range at the back of the FOB near the fuel point just as the sun was coming up. I hoped to be done in thirty minutes and then jump into learning the unit.

I had an easy zero and qualification, and we were just wrapping up when a loud horn sounded. I casually looked around, but SFC Boulis was already grabbing my arm and pulling me toward one of the low concert barriers at the back of the range. He yelled, "Incoming," and we ducked down.

I was just about to ask him about the procedures when a huge explosion detonated less than eighty yards away and the alarm went off. It hit on the other side of a large mud wall, but my ears sure didn't think so. When it was finally safe to get up, we walked twenty yards around the mud wall and found the refueling truck parked there with a helmet-sized hole through the cab. How it did not explode is still a mystery to me. SFC Boulis and I formed an undeniable bond in just twenty-four hours. Combat has a way of doing that.

SFC Boulis was not the last soldier I would have to form a strong bond with in a short amount of time. I believe I could write a novel on just my first thirty days in Baghdad. The 519th was under tremendous operational and internal stress. The unit had logged over 125 wounded (WIAs) and 10 to 15 killed (KIAs) as the unit mounted thirty to forty-five combat patrols a day. The constant training of Iraq police at their active police stations along with continual resupply and refit missions meant the battalion was hit with troops in contact (TIC) almost daily and sometimes two or three TICs. In addition to the enemy, the unit suffered from horrible leadership. The command sergeant major, Mark Green, and executive officer, Dave Detz, were excellent, but the battalion command and operations officer were the most hated men I have ever met. On my first full day, I learned the operations officer was being relieved of duty for being toxic, and the battalion commander was under two separate investigations for toxicity. Dave told me he hadn't told me before I came because he needed me, and the unit needed me badly. *Thanks, Dave,* I thought.

I thought it could not get any more chaotic than taking over as the operations officer in just twenty-four hours with no onboarding, training, or pass-on books, but I was wrong. The summer of 2006 was just prior to the famous surge of additional US forces and the change in tactics. The US was losing two soldiers every day, and dozens were wounded. Our mission to train police was rapidly becoming irrelevant in the deteriorating security environment. We were directed to get the Iraqi police out in the streets if we had to drag them out. This meant we would have to execute a complete change of mission in how we operated, in just a few days. I would rewrite how the battalion would fight within my first week of arriving. I didn't know all the names of the units, had only talked to the battalion commander once, and had only been able to go on patrol two times. My understanding of the environment and enemy was nascent. Luckily, I had some amazing young leaders around me.

In just that same week, sleep had become a distant memory. I had not been this exhausted since my young days at Ranger school. Most nights I crawled into bed at 2200 or 2300 hours, only to be woken at least two or three times every night. Troops in contact, indirect fire, storyboards, medivacs, and closure reports all required my review and approval. By 0600 hours I had amassed little more than four hours of rack and barely could manage to execute some sort of physical training after two cups of coffee. I was finally able to focus on the planning required after downing two Rip It energy drinks. It was within this new state of normal that I attempted to execute what the army called operational design.

The battalion commander, Lieutenant Colonel Bazz, was zero help to either the unit or the mission. During my initial meeting with him, CSM Mark Green had stopped in and stated there was a new pregnant female in the unit. This was not shocking to me, because we had over a thousand young soldiers in close quarters for a year at a time. However, what did stun me was the commander's response. He yelled, "Those willful bitches are trying to get me. They are trying to fuck me over." Before I could even blink, the CSM slammed the door and chewed out the battalion commander.

Mark said, "There are soldiers in that hall who can hear you. You are already under investigation. And I doubt those young women decided to get pregnant to make life hard on you. They are in combat, for God's sake."

The CSM's dressing down seemed to snap the commander back into reality, and he went back to simply talking to me about his priorities and the mission. I didn't hear much after that outburst, as my mind was racing trying to figure out if this man was this vain or suffering from a breakdown caused by the stress of the command. I didn't have time to ponder that question though, as the battalion needed to change its mission.

The plans chief, Captain Karst Brandsma, and battalion S2, Lieutenant Carolyn Bronson, became the nucleus of the planning group that helped craft the way the battalion would pull Iraqi police outside and back onto the streets. Karst was extremely mature and pragmatic and had a solid grasp of the fight, as he had been a company commander in Baghdad already. I had to lean heavily on him as I didn't even know all the names of the units, let alone their locations or combat strengths. Carolyn was only a lieutenant in name. She was an absolute powerhouse of energy and tenacity. Because of my absolute ignorance, I had to listen and learn from Karst and Carolyn. They reminded me that it is critical to listen and support your subordinates, because every unit has troops just like them, and it is a leader's job to give them the ability to make the difference.

No matter how good his troops were, nothing could save the battalion commander from himself. I can best describe the battalion commander in the context of how he responded to this operational shift brief. The team and I had worked almost eighteen hours on the operational design and change of mission concept. I was following doctrine and asked the battalion commander to hear our mission analysis, planning facts and assumptions, and thoughts on the course of action development. He obliged, and I laid out our work in a succinct thirty-minute brief. At the end, he simply said, "S-3, follow me."

I felt two emotions as he sat me down in his office and dragged a dry-erase whiteboard easel into position. On one side I was upset that I had done so badly that he felt the need to draw out what I needed to do, but on the other side, I told myself after only one week I shouldn't expect to get anything right.

However, what the battalion commander drew was a PowerPoint slide. For over ten minutes, he vividly explained how he wanted the storyboard highlighting our new mission set to look. He told me how the other battalion was beating us because they had a professional photographer from one of their National Guard units. He explained the angle he wanted the pictures from, who was to be in the photos, and what patches were to be seen on the sleeves. He even went so far as to ensure we set the camera to high resolution across the entire battalion to ensure good photos.

At first, I thought this was all a joke or some elaborate prank. However, it became obvious he was laser focused on how we looked. I asked him if he had any thoughts or comments about my radical rewrite of our overarching operations set, to which he responded, "Oh that tactical stuff? I know you have got that."

His comment terrified me. He applied all his energy to how to make himself look the best and no energy to whether we had the best plan to complete the mission and care for the troops.

Sergeant Fowler commanding the author's Armored Security
Vehicle as they prepared to enter Adhamiyah

Luckily for the soldiers within the battalion, the commander's narcissism and my tactical ignorance were offset by some of the most professional soldiers I have ever witnessed. I continued to go out as much as I could to try to close the gap between what was occurring in the streets and my limited understanding. During these patrols I began to see true heroes. Our battalion chaplain, Pete Keough, racked up well over a hundred combat patrols, boldly traveling unarmed with multiple squads a week. The company commanders and first sergeants patrolled into Baghdad almost daily, trying to both motivate and train both Iraqis and their own formations at the same time. Every young sergeant was leathered and rigid in their discipline and resilience, while the lieutenants carried the responsibilities and stressors of officers two or three ranks senior. Sergeant Fowler, who was my driver and more importantly my PSO (protective security officer), or what civilians would call my body-guard, was a perfect example. On almost every patrol, he covered my back no matter where I went. From the labyrinth of Sadr City to the alleys of Adhamiyah, he quietly walked with his twelve-gauge pump shotgun if we were in close quarters, checking every corner and assessing every threat without hesitation. He was barely in his twenties, but his demeanor was that of an old vet. In his character, I saw the iron will of our NCO corps and volunteer force.

In just thirty days in the country, my world had transformed into something I sometimes struggle to explain. I existed on massive amounts of caffeine (800 to 1000mg daily) and constant nicotine from dip, chew, and smokes, offset only by IVs and aspirin when the headaches incapacitated me. At the same time, I was bonded to men and women I didn't know before, and now they were my family. The mission, the enemy, and the struggle beyond the T-walls were still unclear to me, but the intensity of daily life gave me purpose. It was this purpose and my resolve that would be put to the test in the coming months.

CHAPTER 13:

YOUNG HEROES AND OLD FAILURES (2006)

❦

Fight hard for what you know is right,
because you must live with it.

As the heat of the summer extended into early fall, so did the insurgent pace outside the wire and the turmoil inside. The Sunni and Shia insurgents had made large swaths of east Baghdad almost impossible to enter without large formations. In addition, explosively formed penetrator (EFP) IEDs, which first showed up in 2005, were now rampant in Baghdad. EFPs had the ability to slice through up-armored Humvees like a hot knife through butter. Paired with remote controls, pressure plates, and elaborate concealment, they were taking a toll on both freedom of movement and morale.

Meanwhile, the command of the 519th continued to deteriorate. Dave Detz returned to Fort Polk to take a new position, leaving me as the only field-grade officer. Lieutenant Colonel Bazz increasingly focused himself on both investigations looking into his actions, one in Iraq and one at Fort Polk, as they intensified. He would spend days disengaged from the daily operations of the unit and generated countless patrols to travel to see his lawyer over on Camp Victory. The battalion headquarters was held together by Command Sergeant Major Mark Green and a very amazing team of NCOs and officers galvanized by their dislike of Lieutenant Colonel Bazz and the love for one another and soldiering. The team jokingly called me B3O when Bazz was not around. It stood for battalion commander, S3, and executive officer, and it was their way of acknowledging I executed all three field-grade jobs at once.

The enemy and Iraqi police themselves did their best to grind us down. We took indirect fire onto our small FOB several times a week, including from the guys I dubbed the Mortar Bandits. These particular shooters would drive past our FOB in one of a thousand gray vans in heavy traffic. In a flash they would stop their van, open the side door, flip out a 60mm mortar tube, launch three or four rounds, flip the tube back in, close the door, and melt back into traffic. Most of the time the van door was shut before the first round hit. I remember one time looking out a window in a small office watching numb as the first few rounds hit at seventy-five meters and then fifty meters. Finally, my brain clicked, and I dove on the floor, along with four others, as the last two rounds hit. We were lying on top of each other like we had just fallen playing Twister. Once the last round hit, we all started giggling at our awkward positioning, until we heard someone scream, "Medic."

Outside the wire, the insurgents had added snipers to match their EFPs. I was out on patrol one day when I heard the radio call stating a soldier was shot in the head by a sniper. As we calculated the route to support the patrol, the squad made a second call stating the soldier was RTD (return to duty), meaning he only had a minor injury. I was furious that this squad falsely reported a headshot and made the entire unit go into hyper mode. I told them I wanted to see the soldier and the patrol leader when they got back to the FOB. When they arrived, I was ready to cut loose on the patrol when their company commander came up and informed me that the soldier had been shot, but it was minor. I asked to see the trooper, and up came a teenager with a huge white gauze on his chin. As he removed the gauze, I was able to see a perfect two-inch-long, quarter-inch-wide groove dug across his chin only two or three millimeters in depth. The bullet that grazed him would have blown his jaw off if it had hit just a fingernail over. I couldn't fathom how close he came to death, and I couldn't fathom his bravery as he mounted right back up the next day and went on patrol.

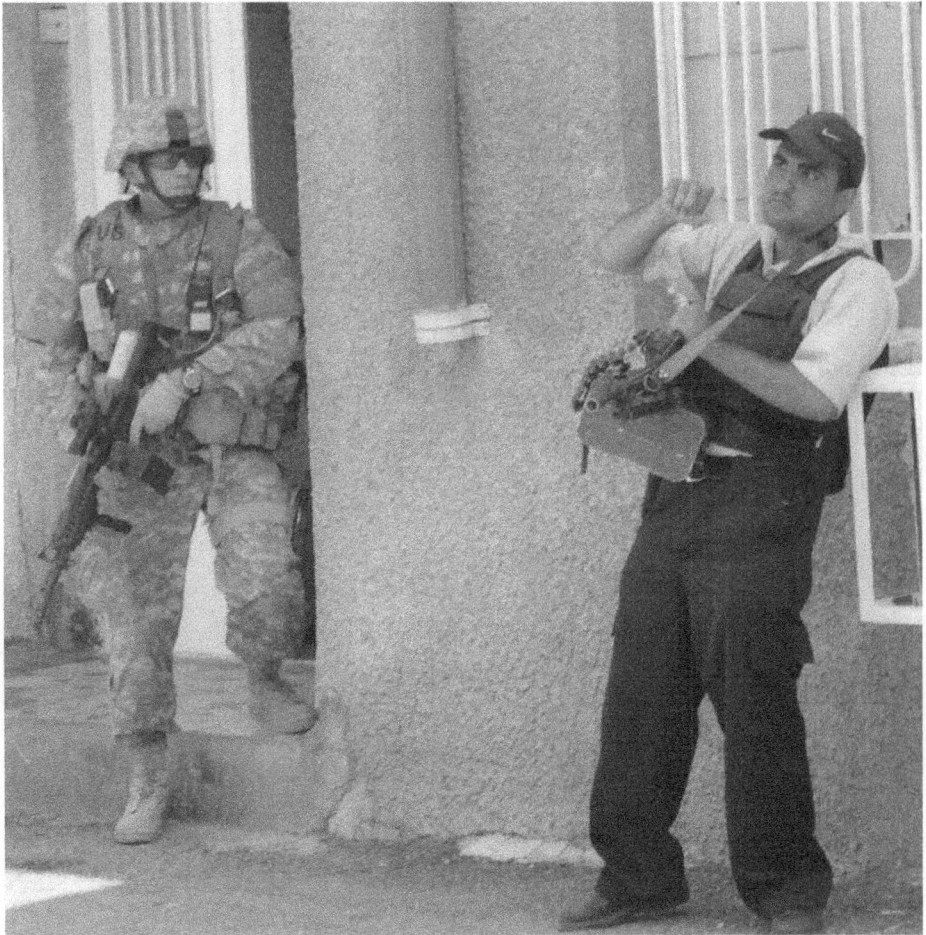

Author on dismounted patrol with Iraqi police September 2006

By September we had a lot of patrols in the streets of Baghdad, and Camp Rustamiyah was overflowing with troops. The surge officially started in January 2007, but we already had crammed two more MP companies and an entire infantry battalion within the walls. Senior leaders at the national level were not ready to push forces out into smaller patrol bases, so our crowded camps became even more appealing targets for rockets, mortars, and artillery. We began taking fire almost daily. It was only a matter of time before a rocket scored a direct hit on a large building on our FOB. One of our newly arrived units, the 118th MP Company, was housed in several large tin-roof buildings. Their commander very adamantly pleaded with me to get them a hardened structure with a solid roof to live in. We quickly added more bunkers, reinforced the interior walls, and tried to harden the structure. However, I could not find a solution to the roof that would not take a year or longer to build. I told him he would have to make the best of what he had; I still feel the sting of my words today.

On the ninth of September, just after breakfast, a single Katyusha punched right through the roof of 118th's barracks and exploded in the rafters, spraying shrapnel throughout the building. There were almost a hundred soldiers in or near the building as morning patrols were prepping to move and night duty troops were in bed. I was in my office when I heard the explosion and then the alarm for MASCAL (mass casualty event). I ran down to the barracks in under a minute, and already dozens of injured and dazed soldiers were being treated and evacuated to the clinic three hundred yards away by almost a hundred responding medics, soldiers, and anyone nearby.

I was attempting to assess the situation when I saw a trooper on a stretcher and a Gator (small UTV) being brought over to load him. I instinctively went over to the group and leaned down to check on the soldier. I did not see any blood or noticeable injury and thought he was just shell-shocked from the explosion. But when he looked at me, I felt an overwhelming sense of fear. His eyes were asking me for help and asking why this had happened. I stammered to tell him he was going to be OK and that help was here. I squeezed his hand as the team lifted him onto the Gator and watched as they drove away rushing to the clinic. I just stood there for a few seconds in shock myself until an NCO yelled, "Clear."

His call snapped me back, and I realized responders were still trying to clear all the rooms and account for soldiers. The response effort was nothing short of stunning. We had medics from every unit on the FOB there and well over one hundred responders treating soldiers, clearing the building, and extinguishing small fires. Within five minutes every injured soldier was transported, and now leaders were trying to account for soldiers, sensitive items, and damages. I had to give very little direction, and shortly after I was led into the smoking building by a platoon sergeant to assess the damage.

The building was dark, and the air was gray with smoke and the unforgettable smell of burnt powder. It is a smell that every soldier instantly recognizes long after they retire. In the roof I could see a small hole about the size of my fist indicating where the round had entered. It then struck one of the roof supports and exploded. Most of the damage was confined to one room housing eight or ten soldiers. I could see the holes punched through the plywood walls, footlockers, and bunks. The pattern was so random. Some areas had one hole, some twenty, and some were untouched. The platoon sergeant said the young man I had helped load on the Gator was lying in his bunk in this room just resting.

Most of the next few hours are a blur to me today. Moving soldiers, repairing the building damage, replacing destroyed air conditioners and power supplies, getting counseling support for the unit, adjusting patrol plans, and just being with the unit. The commander would barely look at me. I believe he blamed me for his soldiers' injuries. I wanted to tell him how hard I tried to solve the problem, but that would have been wrong. And to be honest, I blamed myself as well. I was supposed to protect these young troops, and I had failed. Lieutenant Colonel Bazz was nowhere to be found, so I did the best I could to keep the team focused on recovering until the clinic called.

PFC Anthony (Tony) Seig, the soldier, I had just helped a few hours ago, had died. He was just nineteen. He had joined the army knowing he would see combat. He had stepped forward to serve his nation when others just sat home and talked about patriotism. He was one of the good guys. We failed him, but he never failed us. I still see his face, and I still wish it was not true. I didn't realize it then, but Tony was going to play a large part in how I would try to lead going forward. However, that day I suppressed every emotion I had and focused on doing the next task that needed to be done.

The tasks were coming faster now than I could process. I began to see that the army's brigade combat team (BCT) concept put too much responsibility on a single commander and staff, leading to countless issues and challenges. BCTs, which were already large with almost 3,500 assigned personnel, swelled with attachments, contractors, and joint augmentees to as large as 5,000. As one of the functional battalions, the 519th operated in several BCT battle spaces but was not directly under their command. However, we had to support their operations, attempt to gain their support, and most significantly, live with their actions, good and bad. The few BCT commanders I met presurge believed they could kill their way to victory and saw training Iraqi police and military as a distractor to be tolerated only.

Currahee 6, the callsign of the BCT commander who controlled most of eastern Baghdad, was one such officer. Every interaction our unit had with him or his staff was poor, both in conduct and analysis. I do not know whether the commander was as incapable as he appeared to our unit or was just too overwhelmed to function at the level he found himself. However, I do know we lost synergy, ground to the enemy, trust from our allies, and soldiers because of this disfunction. His arrogance was palpable, and his disdain for alternate thoughts matched that of the 519th commander, making a deadly combination for soldiers. He once called my phone directly the first month I had arrived, when he felt one of our patrols was not doing something correctly, and said, "What the fuck are you doing over there? I know you are a dumb motherfucker, so go put your XO on the phone now!" I had never met him before that phone call, but in that ten seconds, I learned a lot about his character.

Currahee 6's attitude permeated throughout his headquarters. Staff and subordinate units were scared to engage him and quick to insult, bully, and push off problems. I did not realize how dysfunctional that unit was until I worked with several other BCTs facing the same stressors but with significantly more synergy, esprit de corps, and motivation. It was because of this toxic culture that I was not surprised to find orders that made poor sense went unchallenged or revised. One of those orders was brought to me by Captain Shevin Denmark late one evening.

Shevin's unit had been tasked to check on the police checkpoint daily to ensure it was resourced and executing its mission correctly. This was not an unusual task. In fact, surrounding Baghdad with a ring of checkpoints to prevent the infiltration of insurgents, explosives, and weapons was central to the strategy of protecting the new Iraqi government. However, what was unusual was the use of the term "at." The order we had given Shevin stated he had to check this checkpoint at an exact time daily, meaning he was predictable. And as Shevin pointed out, this point was exacerbated by the fact that there was only one way to get to the checkpoint, and it ran parallel to a canal. This meant his troops had to travel the same road at the same time of day every day. It didn't take much for even me to see that this was dumb. My mistake was thinking it would be easy to adjust it.

I had my staff investigate the origin of the order, and they determined the order had come from Currahee. Technically, the BCT could not issue direct orders of this nature to our battalion, but they did when it came to the police, and historically the battalion executed them. I asked my operations team to work with the brigade to get more variation in the timing and frequency of the checkpoint visits to allow the units to remain hard to pattern by the enemy. The request was flatly rejected, and my emails and calls to the brigade operations officer went unanswered. So I informed Lieutenant Colonel Bazz, on one of the rare times by now that he was around the headquarters, of my intent to stop executing the mission. I didn't even get to finish my reasoning before he cut me off and told me we would not stop executing. I asked why, and he said Currahee 6 had been good to him. I had been stopped in my tracks and issued my marching orders, so I bit my lip, shook my head, and moved on. Unfortunately, the enemy did not move on.

About two weeks later, I was heading to the chow hall for dinner when my ICOM (handheld secured radio) crackled to life and asked me to return to the TOC immediately. As I arrived, the current ops desk was quickly receiving information and issuing support orders for reinforcements and medical evacuation. Before I could ask, the battle captain said the patrol ordered to check that checkpoint at 1700 hours had been hit with a complex ambush consisting of five EFP detonations. In an instant half the squad was injured, with one of the four vehicles absorbing three hits itself. At least two soldiers were severely wounded from that truck, and forces were rushing to the scene.

Over the next four hours, the battalion went through our well-rehearsed drills of troops in contact. The staff was so hardened, efficient, and trained that again I did not have to lift a finger. Medical evacuation, reinforcement, recovery of equipment, and retrograde of personnel all were executed flawlessly. The wounded were airlifted, critical reports sent, and site exploitation begun. At 2100 hours the heavily damaged Humvee was dragged back into the FOB. As the S3 I handled the mission. Since I was also the XO, I had to handle the recovery operations as well. Within our battalion, senior NCOs and officers did the cleanup of damaged vehicles. We did not want young soldiers or any soldier from the impacted unit to experience it.

When the truck arrived, I stepped behind the tent flap at the motor pool and stared at the mangled frame. I opened the passenger door and was overcome with the smell of "pennies." There was dried flesh and blood throughout the space. I stared at the dried bloody fingerprints on the radio buttons and imagined the vehicle commander trying to work his radio as he bled. The strong smell of copper came from both the blood and the copper metal used in the projectiles that had penetrated the truck. The Humvee had been hit with three projectiles, each about the size of a quart of paint. The insurgents angled them to hit the engine, gunner, and passenger at the same time. Once they penetrated, the copper spalled into fragments that passed through human flesh and embedded into the interior. Gatorade bottles, personal bags, weapons, food wrappers, and ammunition were all painted in dark red and copper. I had failed to prevent this predictable attack and the stupidity that allowed it. I am reminded every time I hold a penny.

CHAPTER 14:
COMING HOME CHANGED (2006–2007)

❦

You can leave the war in a day, but it will not leave you the same.

GETTING OUT OF IRAQ and home to Fort Polk was almost as hard as getting into Iraq. By October 2006, units were streaming in to begin surge operations in January. We were executing daily missions, adding new units, attempting to hand off to our replacement battalion, and packing at the same time. Because of the way units rotated in and out of Iraq, our headquarters unit and two of the organic companies would not depart at the same time. It would take almost sixty days to get everyone from the 519th back to Fort Polk, Louisiana.

After a simple ceremony on a dirty motor pool parking lot, our headquarters company mounted helicopters just past midnight departing the hell of Rustamiyah. It was weird to sit at Baghdad International Airport (BIAP) with unloaded weapons and nothing to do but wait. When we landed in Kuwait the next day, we learned a volcano erupting near Iceland had canceled most of the flights across the Atlantic, and we would be stuck in Kuwait sitting in tents for several more days. I just laughed. Who else gets to have a volcano impact their tour in Iraq?

During those three days, it was emotionally exhausting for everyone. The continuous high of combat was replaced by nothing. People crashed into endless sleep, withdrew into reflection, or were euphoric about leaving. The combination of these three subsets made for awkward moments within our crammed tents. Lieutenant Colonel Bazz completely withdrew from the moment. Command Sergeant Major Green told me Lieutenant Colonel Bazz knew something was going to happen to him when he returned to Fort Polk based on the investigations, but he did not elaborate. Almost all the soldiers avoided Lieutenant Colonel Bazz at all costs. Chaplain Keogh, Command Sergeant Major Green, and I agreed to attempt to include Lieutenant Colonel Bazz in trips to the chow hall. I did not like him, but I felt empathy and pity for him. He had led a massive battalion for a year in a tremendous fight, and now his mission and most likely his career was over. He had no family, no friends that I knew of, and now no purpose. He was broken.

Unlike the depressing desert of Kuwait, the flight home, arrival in the US, and welcome at Fort Polk were amazing. The chartered plane had enough room for us to spread out and sleep. The USO volunteers cheered us in New England when we landed to refuel and fed us all types of snacks and sweets. Once we touched down in Louisiana, police and fire trucks escorted us, and signs lined the road. It was special, it counted, and it cannot go unmentioned how much it meant to us. When we finally formed up and marched into the gymnasium for our welcome home reunion with our families, the mood was incredible. I scanned the crowd, trying not to violate my military bearing until I saw Angela and Hunter waving American flags. Hunter was dressed in an army uniform, grinning from ear to ear, and Angela was like a dream. When I saw them, I knew I was home after more than fifteen months and a world away.

As Lieutenant Colonel Bazz saluted the brigade commander and declared the battalion was home, I was champing at the bit to let go of the army for a few weeks and spend time with my family. I was still daydreaming when the brigade commander told Lieutenant Colonel Bazz to follow him and then turned to me and said, "Take charge of the battalion."

What the hell? That was all I could think of. Did I miss some part of the ceremony in my mind? I stepped up to the commander position and did the only thing I knew to do. I waited until the brigade and battalion commander departed, then I turned around and yelled, "Fall out!"

The best hug ever

In an instant, the gym became a mad rush of families and friends colliding with their soldiers. I scooped up Hunter and kissed and hugged Angela, and then we all started to cry tears of joy. I introduced them to my new blood brothers and sisters all around us. It was clear that there was a bond here, much different from just a normal unit. After nearly fifteen or twenty minutes, we realized no one had departed. It was almost like we could not believe we were going to separate from one another. Meanwhile, Dave Dietz, my buddy major who had departed early to come home and become the brigade executive officer, was waiting for me on the side while welcoming soldiers home himself. I made my way over to him and asked where Lieutenant Colonel Bazz was going and when he would be back.

Dave looked at me and said, "He is not coming back. He is being relieved of command, and you are now in command. The army was working to get a replacement here, but until then we will have to make do."

I realized my break would be short lived, I would not be able to disengage like I had hoped.

The next two weeks were surreal. I found myself all over the place mentally. I was so happy to be home, but I would zone out or have trouble just being still. America was not at war and Louisiana was a safe place to be, but I could not turn my mind off. I was constantly thinking, constantly evaluating my surroundings, and looking for trouble. I didn't realize it then, but I was quick to anger when simple things failed to work. Hard stuff intrigued me, but simple issues or failures on my part were met with immediate anger. I was suffering signs of something I didn't recognize or understand. Looking back, I was able to power through because of my amazing wife, my desire to make up for lost time with Hunter, and the mission of continuing to lead the battalion.

As soon as I returned to work, I was quickly operating at deployment speed. The army had placed a lieutenant colonel on orders as the temporary battalion commander until one could arrive in five months. However, this officer had another job and simply was there if I needed an O-5 to sign anything. This left the day-to-day operations up to me. I was doing the jobs of three field-grade officers at once and trying to fix an absolute mess left by the void of leadership. Command Sergeant Major Mark Green was helping greatly but had decided to retire rapidly after this combat tour, and his loss as my wingman added to my stress.

It is hard to articulate what rough shape the unit was in and what monumental efforts were made by young officers and NCOs to fix it. As an example, I had to sign evaluation reports for dozens and dozens of people I barely knew. The loss of the battalion commander, previous executive officer, and operations officer meant many leaders would not get evaluations of their critical combat experience if I did not fill the void. Additionally, the battalion had dozens of lost equipment investigations that involved captains, which meant I had to do them all as the only senior officer. We had to reset all our equipment, consisting of a hundred thousand pieces, and still safely bring home over three hundred more soldiers from the fight.

The bureaucratic challenges paled in contrast to the challenges of caring for our soldiers. Our Wounded Warrior roster, the tracking system that managed soldiers in the process of medical discharge, swelled to over 150. Almost 20 percent of the battalion was dealing with various types of issues from the deployment. Mild traumatic brain injury, PTSD, and severe depression accounted for a large portion of these soldiers. Each one of their cases was unique, and as the units returned and split up, the risks of failing these soldiers went up. I was constantly torn emotionally between getting the 519th back up to combat readiness as fast as I could for the army and caring for each soldier as they deserved. I knew I was failing at both tasks, but I just hoped I was not dropping any glass balls.

The young officers and NCOs who were still on the battalion staff or serving as command teams were as committed as any I had ever seen. Karst Brandsma, Dave Charbonneau, John Petkovich, Vinnie Particini, Carolyn Bronson, Pete Keough, Dave Koch, Duran Davis, and a host of others never missed a beat when it came to their dedication to soldiers and the army. They simply gave all they had every day to both the army and one another. If it were not for their dedication and love, the 519th would have fallen into disarray and languished there for years. Instead, by April 2007, the battalion was beginning to scab over the wounds of Iraq and focus on the next mission. I was finally finding some peace and enjoyment again at work. I still struggled with the events of the last year, but I believed I would continue to improve. The deaths of my father, Jeff Toczylowski, Rex Sprague, Tony Seig, and dozens of wounded were a lot to process, and I knew this. The chaos of war, the waste, and the heartache were just going to need time, I told myself.

It was during this time that I met an amazing new role model. Lieutenant Colonel Brad Graul arrived at the 519th about five months after the battalion returned from Iraq. He had been pulled off the alternate command list and urgently directed to Fort Polk. Immediately you could tell he was going to be the leader the battalion needed at this time. He was a quiet, common-sense man with well over two decades of service. He was a man of faith, strong values, and absolutely no ambition to climb the rank ladder. He had to leave his family at another duty station, so Angela, Hunter, and I quickly became his extended Fort Polk family. Like Lieutenant Colonel Spain before him, Lieutenant Colonel Graul loved to be around soldiers. He was capable of just talking to them about sports, food, and family. His favorite line was to invite soldiers to eat breakfast with him in the chow hall and drink chocolate milk. Soldiers have a tremendous skill to sniff out a fake leader who claims one thing and does another. So when you saw soldiers gravitate toward Lieutenant Colonel Graul as he traveled around the unit, you could tell he was genuine.

Shortly after Lieutenant Colonel Graul arrived, both a new S3 and CSM arrived as well. My workload dropped 75 percent, and I was once again passionate about coming to work every day. I was getting home for dinner and spending hours of fun with both Angela and Hunter. I was finally in a great groove and barely noticed the events playing out at the next level of command. The brigade we were a part of was transitioning from a provisional brigade (temporary construct) to a newly designed maneuver enhancement brigade. The commander who had relieved the previous battalion commander, Colonel John Moore, was leaving, and Colonel Martin was taking his spot. What I did not know at the time was that Colonel Martin would become the closest thing to the devil I would meet.

CHAPTER 15:
MUTINY AT TIME OF WAR (2007–2008)

❦

Surrounding yourself with people of high moral character
ensures you measure up when needed.

IN MID-MAY 2007, I had just completed a great day at work when I happened to see a Facebook message from an old college ROTC buddy. I was excited to open the note and see what he was up to. However, as I read the words he wrote, my mind went blank, my face went numb, and I began to cry. I wanted to run, and I wanted to fight at the same time; I was emotionally smashed. That was how I found out that Larry Bauguess, one of my first peer mentors, was killed in Afghanistan. He was killed by a coward masquerading as a friendly Afghan policeman. We kept losing the very best among us, and no one beyond the military seemed to understand that it was just not fair.

I was still emotionally withdrawn in June when Lieutenant Colonel Graul told me the new brigade commander had selected me to be his executive officer. This was normally a great honor, where the top major of the brigade was selected for this position. But I was not interested in taking the position. The loss of Larry had galvanized my desire to stay with the soldiers of the 519th and prepare them for their next deployment. When Lieutenant Colonel Graul talked to the brigade commander about what I wanted, he came back and told me, "Glenn you better go. Colonel Martin said you can either be number one of thirty majors or number thirty of thirty. He will not be told no."

I reported in July, along with another major Steve Basso, who was to become the brigade S3 and one of my lifelong brothers. When we arrived, the brigade was only one by name only. The headquarters only had 20 percent of the staff, a fraction of the equipment, and no permanent headquarters. Steve and I were faced with an activation ceremony, headquarters move and build out, and a certification exercise, all before November and all while attempting to receive over a hundred new officers and NCOs. But before we could even begin to knock down all the tasks before us, we had to meet our new commander, the one who had cherry-picked us up to his level.

The first day I met Colonel Martin, I will never forget. The first impression I had was shock that he was an infantryman. He was medium height, but clearly thirty or forty pounds overweight. His face was flush, and his breathing was labored. I knew he had just finished a deployment in Iraq on an adviser team, so I thought maybe the pace and stress of the deployment had taken an extreme toll on him. He welcomed us in while sipping an energy drink. There was about five or ten minutes of pleasantries, and then he blatantly said, "Not all majors are created the same. You two will be numbers one and two in the brigade and run the unit for me."

I remember recounting the meeting to Angela when I got home. She asked how his statements made me feel. I told her I felt like I just joined the mafia, like he was trying to buy my loyalty. I had never had an army leader who had not worked with me yet had started off by telling me I was going to be ranked number one. The little voice in my head told me to be careful. But I was still a young major who had not worked closely with colonels before, so I thought maybe this was how senior officers functioned. I didn't spend too much time pondering his motives after that first meeting, because the to-do list for the new brigade was monstrous.

Over the next several months, Colonel Martin was true to his statements for the most part. He was very hands off in the daily operation of the brigade. Steve and I had carte blanche to issue orders, guidance, and direction across the four battalions and staff. He rarely left his office and did not visit subordinate units or troops. He seemed to focus his energies on two things, getting personnel for the new brigade and finding a deployment for the brigade to execute. He went to great lengths to get personnel for the unit, including frocking (early presentation of a promotion rank) the new personnel officer to major and sending her and her staff to Human Resource Command to lobby for new personnel. On the top of the list, he told them, was to secure Staff Sergeant Wheel's move to the brigade. Colonel Martin had worked with her in Iraq, and he said she was the best intelligence analyst he had ever seen.

During these early months in the brigade, I worked twelve or more hours a day with Colonel Martin. We both shared a passion for deer hunting and guns and would even go deer scouting during a quick lunch break from time to time. Angela and Hunter would come to visit me at work, so they got to know Colonel Martin as well. I noticed that he liked to run the brigade using just me and Steve as conduits. He rarely met with battalion commanders and was very comfortable letting them run their battalions with little or no oversight. I was consumed with tactical details daily, having great success building the unit, and I had a great relationship with my boss. I thought he was different from any combat arms commander I had met, but I didn't see any major red flags. This all changed the day Staff Sergeant Wheel arrived at the unit.

I had all but forgotten about Colonel Martin's interest in getting Staff Sergeant Wheel to the unit when she appeared in the command section of the headquarters. She was dressed in civilian clothes with her hair down past her waist and had her young baby at her side. The secretary ushered her straight into the commander's office. Then, within a minute, Colonel Martin called me in and introduced me to her. As I was walking in, the little child hugged Colonel Martin and called him "Poppy." As I stepped back to my office, I tried to rationalize what I had seen. I thought Colonel Martin must have been really close to his small team downrange and was seen as a grandfather figure, but that familiarity might cause problems in a line unit.

As the executive officer of the brigade, I was uniquely involved in almost all aspects of the day-to-day operations. This also meant that I was uniquely positioned to realize there was more to the relationship between Colonel Martin and Staff Sergeant Wheel that others could not see. I saw Colonel Martin direct the diversion of an intel Master Sergeant, even though we were critically short of this type of NCO. This diversion meant Staff Sergeant Wheel would continue to serve in a position two grades above her rank, which was great for her promotion potential. I watched Colonel Martin direct our newly arrived government cell phones to go to staff primary officers and their senior noncommissioned officers before lesser staff sections. I realized this meant Staff Sergeant Wheel would have her own government cell. And I witnessed Colonel Martin respond to trivial training issues that only Staff Sergeant Wheel could have informed him of. I was beginning to see that Colonel Martin was far too involved in Staff Sergeant Wheel's daily actions.

If these initial oddities were not concerning enough, there were three successive incidents that left no doubt in my mind that things were wrong. The first event occurred just a month before we deployed for our certification exercise. I had the brigade staff in a battle staff training exercise trying to figure out how we were going to fight this new formation without any history, concepts, or standardized operating procedures. It was a grueling and stressful event for the command. As we began, I directed the guards at the front of the command post to tighten up their procedures and told them how I wanted them to control access. I then jumped right into the hard work at hand. However, in just a few minutes, Staff Sergeant Wheel came up to me in front of the entire staff and asked who had given me the authority to direct "her" soldiers.

I was not in the mood to be challenged in public as we faced the incredible task of building a battle staff for the validation exercise in just weeks. I cut her off hard, told her she was confused about both her authority and mine. I then said, "Move out, Sergeant, and focus on the tasks I gave you." I immediately thought this was the effect of her close relationship with Colonel Martin and her immaturity. She was confused about how a line unit operated, and she would be in for a steep learning curve.

But I was the one in shock when Colonel Martin arrived later that day. The first thing he said to me was he heard I had an issue with Staff Sergeant Martin. I was surprised he was tracking such a minor issue, but I told him it was just a young NCO confused about unit operations, and I would handle her. He raised his eyebrow, and I asked if there was an issue. He simply said, "Not yet."

The second incident occurred just after we moved into our new headquarters late one evening as we were preparing to depart for the night. Steve Basso, the S3, and I were standing in the command group with Colonel Martin going over the major events of the day when Staff Sergeant Wheel came in to deliver something to Colonel Martin. Colonel Martin commented to Staff Sergeant Wheel about why the security section put cypher locks on the command group door when there was a glass window in the door that one could break and climb through. Staff Sergeant Wheel just blurted out, there was no way for someone to fit through the window if they were as big as he was.

I just froze in shock. Staff Sergeant Wheel had just called the brigade commander fat to his face in front of two majors. I had never seen something so unprofessional, disrespectful, and a clear violation of discipline. I immediately glanced over at Major Basso, and he looked back with the same shock on his face. Then out of instinct we turned to the commander prepared to execute whatever orders he issued toward this blatant comment. But to our amazement, Colonel Martin just laughed and said, "Someone is sensitive about the doors," and then he changed the subject. I believe it was at that second that I knew their relationship was beyond mentor and mentee.

The third incident was our unit's readiness exercise. About a hundred brigade staff headed to Fort Leonard Wood, Missouri, to fight a battle simulation for a week. At the end, the First MEB would be certified as fully mission capable. The exercise alone was complex enough since there was no doctrine for how this new unit was supposed to fight. However, onboarding new personnel, creating both the physical and operational systems, and transporting them all over two hundred miles away made it even more difficult. But as difficult as the mission was, it helped to fully unmask the brigade commander's true motivations.

As we prepared to depart for the exercise, the aloof commander unexpectedly inserted himself in trivial details. He directed us to stay in hotels off post, directed who stayed at what hotel, and specifically stated Staff Sergeant Wheel needed to travel in his vehicle and stay in the same hotel as him since she would transport the classified documents. When I told him there was no need for her to travel with him, since all staff had classified courier cards, he exploded and told me not to take away her job. Even our newly arrived deputy commander, Lieutenant Colonel Jesselink, found the brigade commander's actions incomprehensible.

If Colonel Martin's actions leading up to the mission were unusual, his actions during the trip were even more incredible. In just one example, halfway through the exercise, Colonel Martin stated he wanted the senior leaders to go out to dinner together. I asked if he wanted the field-grade officers or senior NCOs to come, and he stated he would keep it small. So Steve Basso and I traveled together to the designated restaurant, and when we entered, we saw the deputy commander, command sergeant major, the colonel, and Staff Sergeant Wheel. We then looked at one another, and we were dumbfounded. For the entire dinner, Colonel Martin talked openly like he had an executive counsel around him. Meanwhile, Staff Sergeant Wheel sat there beside him enthralled in it all.

As we concluded the exercise and unit activation and moved into our new headquarters, the pace slowed down just enough at the beginning of December for Lieutenant Colonel Jesselink and me to finally talk holistically about what we were seeing regarding Colonel Martin. By then, I had assessed Lieutenant Colonel Jesselink as an officer of high integrity and selflessness. As we discussed what we were seeing, we were still unsure how we should approach Colonel Martin and the growing perception within the brigade. He was going more erratic. He had just blown up about the installation turning off our text ability on our government cell phones. Texting was a new concept then, and most of us didn't even know how to do it. But I begged the installation deputy commander to turn Colonel Martin's back on because he said he was communicating with several generals this way. I thought that was the last issue before the holiday. However, fate and dumb luck would force our hands in the most unlikely way.

It was two weeks before Christmas 2007, and the brigade had begun a half-day work schedule for the holidays. I headed off at lunch to the local cellular store in search of my wife's Christmas gift—she wanted the new Razr phone. However, they told me the best deals were online, so I decided to head back to the office and look there. When I got to the website, it asked for a code. So I called the cellular company, and they asked if I had a cellular phone from them so that they could send a code to it. I said yes, but it was a government one. They said it didn't matter, and they sent the code via text. I entered the code into their website and immediately gained access to search for cell phone deals.

However, the page that popped up when I entered the code happened to be the complete account for all government cell phones at Fort Polk, Louisiana. In a weird glitch, I had access to everyone.

However, the drop-down menu had defaulted to my number, and I was staring at two graphs: one for voice and one for data. The voice was green, but the data said Not Authorized. I immediately thought I had better check the commander's graphs because he had his data turned back on and could be incurring illegal costs. As I selected his number, the data graph was red. I saw a plain-looking hyperlink right below the graph that merely said Details. So I clicked on it hoping to understand what he was being charged for.

In a flash, the screen filled with a spreadsheet showing numbers the phone was texting, and the times of those texts. Even though the sheet did not say what was in the text, the immediate detail I noted was that almost all of the more than two hundred texts in the last ten days went to one number at all times of the night and day. The number was local, and as I pondered this, I remembered his direct involvement in the assignment of government cell phones three months prior. I turned to my right and looked directly at the contact roster I had on my bulletin board. There it was—Staff Sergeant Wheel's cellular number. I stopped breathing and started sweating. I had just stumbled onto the smoking gun. I printed the pages, placed them in a folder, and left the headquarters as quickly as I could.

Angela and I talked about what this meant in earnest all night. The next day, I briefed Lieutenant Colonel Jesselink, my direct senior, on what I had discovered. I can still hear his initial response ringing in my ears: "Agh, Glenn, we just shit and fell back in it now."

We both agreed to reach out to some gray beards (a military term for mentors who are no longer in your chain of command or retired) that night and reconvene the next day. I first reached out to Lieutenant Colonel Graul, my previous battalion commander, and he crystallized the issue for me. He said, "Glenn, there is nothing to ponder here. Right is right, and wrong is wrong. Colonel Martin is wrong." I then reached out to Colonel (R) Spain, and he echoed Lieutenant Colonel Graul but also added that we should confront

Colonel Martin and give him a chance to go to his boss first to self-report before we went.

Lieutenant Colonel Jesselink got the same advice from his mentors, and we looked for a time to talk to the commander. Lieutenant Colonel Jesselink said he would look to talk to him alone because Lieutenant Colonel Jesselink was not getting promoted, and Colonel Martin could not professionally hurt him. Again, our hand was forced when Colonel Martin out of the blue made a veiled threat in a meeting about not sending me to a school I had to have the coming year for promotion. What we did not know at the time was that the command group secretary was part of Colonel Martin's coverup and tracking our every move. Lieutenant Colonel Jesselink felt he had no choice but to stay after the meeting and engage the commander.

In the meeting, Colonel Martin admitted it looked bad and asked what he should do. Lieutenant Colonel Jesselink told him he needed to go to the commanding general and ask for an outside investigation to disprove the perception, or he needed to admit to the relationship to the general. Lieutenant Colonel Jesselink left the meeting believing the commander was going to do just that. However, when we arrived at work the next week, it was clear Colonel Martin had other plans. He announced Lieutenant Colonel Jesselink was going to deploy early, I was moving to be the officer in charge of the rear detachment, and the brigade CSM was leaving the brigade. In one stroke he was attempting to remove every senior leader that he believed could expose his actions.

I met with Lieutenant Colonel Jesselink later that day, and we decided we had to go to the commanding general ourselves. We had demonstrated loyalty and respect to Colonel Martin by giving him a chance to self-report his transgressions. However, he made a calculated decision to fight his way out of his troubles by using his rank and authority.

The next day Lieutenant Colonel Jesselink was scheduled for oral surgery, and I was on leave taking my son to his first dental appointment, so we agreed to meet again on Wednesday and plan to see the general. So I was totally confused when I exited the dentist's office and saw six missed calls. As I listened to the messages, my concern spiked. The messages said the brigade commander was looking for me, and I was ordered to report to his office immediately.

Angela was terrified and told me I shouldn't go in. She had heard talk about him being vindictive and violent. She felt strongly that it was a trap. I told her I had no choice in the matter, he had issued me a direct order. As fate would have it, just as we were having that conversation, we were driving by an Office Depot store. I hit the brakes and turned into the store. I came out five minutes later with a micro digital recorder. We spent the remaining forty-five-minute drive back to the post figuring out how to use it. There were no smartphones or video APPs in those days, so this was high tech. During the drive, Lieutenant Colonel Jesselink called me and told me Colonel Martin had fired him and was looking for me. I did not feel good about "wearing a wire," but both Angela and I were genuinely scared of what was in store for me.

As I entered the command group with the recorder in my right shoulder pocket on my uniform, Colonel Martin was standing there waiting for me. He asked how the dentist had gone and asked about Angela as well. Then he motioned to his office, and we went in. As soon as the door shut, Colonel Martin's entire demeanor changed. He told me to sit down at his small conference table and read a counseling statement placed there. I read that I was being charged with mutiny and sedition at a time of war. I struggled to process what I was reading. I focused on another sentence that said the maximum punishment was life in prison or death. By the time I finished that short paragraph, Colonel Martin could tell I was stunned, and he began his verbal assault.

He told me he had already conferred with the general, and he (Brigadier General Bolger) was going to allow Colonel Martin to handle prosecution at his level. Before I could even fathom what he was saying, he told me to stand up and go to the computer. Once I got there, he ordered me to access the phone records and show me how and what I had gotten from the cell phone portal.

I never reached toward the keyboard. I slowly said, "Sir, I believe you are attempting to use your professional power for personal gain. I request we adjourn to the general's office." As I said these words, I was praying that the digital recorder in my pocket was working. I knew both Colonel Martin and I had just crossed a line that there was no coming back from.

Colonel Martin yelled, "We are not going anywhere. Sit back at the table."

Once I sat back down, he launched an over three-hour verbal assault attempting to break me. He went from threats of jail to bribing me with a good evaluation to then blaming me for his failures. He bounced between each attempt like a madman. I was scared beyond any combat situation I had been in. I didn't know if he was telling the truth and the general was on Colonel Martin's side. I had never refused an order in almost two decades, and now Colonel Martin's attack was relentless. Angela was so scared when she had not heard from me after two hours that she contacted Lieutenant Colonel Graul, and he was waiting outside in the parking lot making sure I was safe.

As we reached the third hour, Colonel Martin was going unhinged. He told me if he had found me that morning, he would have put two bullets in my heart and one in my head. As I started to take notes, he reached across the table and tore my notebook up. I slipped my knife out of my pocket and held it open under the table; Colonel Martin was beginning to get aggressive.

He then said he was going to deploy me where I could not protect Angela and Hunter. Hearing him threaten my wife and son triggered me. I looked him dead in the eyes and told him I would kill any man who threatened my family. He said I needed to remember I was a major and he was a colonel. I told him he needed to realize I was a husband and father above all else. At this moment, thank God, he blinked and sat back in his chair.

Finally, after almost three full hours, Colonel Martin realized he could not break me or get me to give him any information on the evidence I had of his actions. He told me I was under house arrest, and I could not communicate with anyone until such time as I was court-martialed. He then ordered me to write my own OER and submit it to him the next day.

I left the headquarters completely shattered. I didn't know whom I could trust or what was going to happen the next day. As I drove home, I pulled out the micro recorder and checked to see if it had worked. It had captured every word crystal clear.

Once I got home, Angela and I quickly formulated a plan. We decided the next day she would leave with Hunter and travel to her parents' house. I made four digital copies of the meeting in Colonel Martin's office and gave them to her. I told her to give them to four family or friends not in the army in case something happened to me. I knew I had to go into the office in the morning and provide my evaluation support form, but not a draft evaluation. That way, Colonel Martin could not say I violated an order. A part of me believed Colonel Martin was just an evil individual, but a part of me didn't know what could happen. I sat up that night with my twelve-gauge by the door.

The next morning Colonel Martin was hunting for me before physical training started. He yelled at me in front of several officers and asked where my evaluation was. He told me to submit it now and go back to my house. I left the headquarters and headed toward the post headquarters. On my way I called Lieutenant Colonel Jesselink. He was fully committed to going with me, but his mouth surgery left him completely incapacitated. He asked if we could wait until the next day, but I quickly brushed that option aside. Then I called Lieutenant Colonel Graul, and he said he would come with me. I thanked him but told him this was my fight, and he needed to protect the battalion from whatever happened.

I entered the command group like I had done many times before for meetings. But this time when I told his secretary that I needed to see the general, she looked sad. She said she was sorry, but she didn't see me on the calendar. I told her it was on the open-door policy that was a standard policy most commanders had allowing soldiers to bring problems straight to the commander. She asked if I had talked to my brigade commander, and I told her my brigade commander was the problem. She finally relented and told me to have a seat.

Brigadier General Bogler was a figure you never missed in a crowd. He was at least six foot six, thin, and liked to drive himself around in an old Ford Taurus instead of being driven in a government vehicle. Every time I had been in a meeting with him or shown him training in the field, he had been soft spoken, reserved, and even tempered. However, I could not help but shake as his secretary told me to go in and he would see me. As I walked in, he beckoned me over to a small sitting area and asked what was wrong. I stammered that I was more scared now than in combat. Before I could say another word, he said, "Do you think Colonel Martin is having an inappropriate relationship with a soldier?"

I was completely stunned by his comment. My mind raced as to what this meant. Did Colonel Martin really have his blessing to charge me with mutiny? Or had a rumor of Colonel Martin's actions already made it to the general's level? Either way, it did not matter at this point. I began to lay out all the information I had to the commander, including the fact that I had the previous night on tape. At that point he told me not to mention the tape in case there was a court-martial. I wasn't sure whether he meant me or Colonel Martin. He told me things were going to happen fast, and I was to go home and remain there until I was called. I left his office feeling like he had given me the same order Colonel Martin did, but hoping his motives were in keeping with the army values I cherished.

My stress level over the next few days was almost unbearable. I could not talk to anyone within the unit, and rumors were flying around that both Colonel Martin and I had been removed for illegal actions. Overnight I was stripped of my close professional support group. But luckily, I had some amazing previous bosses, peers, and senior NCOs who checked in with me. However, my greatest strength came from Angela. At one point she said, "If we get out and you have to go to work at Walmart, I will start a spouse support group there to support your career." Her love, fierce loyalty, and steadfast conviction to right gave me strength.

As promised, I did not have to wait long for something to happen. I was contacted by the garrison commander, Colonel Saige, and informed he would be conducting the investigation of Colonel Martin. This was my first indication that Brigadier General Bogler was taking this extremely seriously. I knew Colonel Martin and Colonel Saige were not buddies by any stretch of the imagination. Colonel Saige told me to prepare a statement and report to see him the next day. The next day I spent over two hours covering what I knew and submitted an eight-page sworn statement.

The following day he called me and asked if I had a tape of the meeting with Colonel Martin. When I said yes, he asked why I had not told him before. I told him what Brigadier General Bogler said and that I interpreted that as not to mention it. Colonel Saige told me to bring all copies of the tape that I had the next day. I wasn't sure how to take his comment, but I brought "all copies I had." When he asked if these were all the copies, I told him that my wife had sent off other copies. He sat back in his chair and smiled from ear to ear. What I didn't know at the time was that Brigadier General Bolger was getting pressure from a specific senior officer to make the case go away. Brigadier General Bolger was committed to doing what was right, and now Colonel Saige could assure him it could not be swept away.

Colonel Martin was suspended from his command, his weapons were removed from his house, and within a few weeks, the investigation was complete. Though the investigation never got the actual text messages, they were able to prove that from June to December, Colonel Martin and Staff Sergeant Wheel had talked on the phone for over twenty-five hours and sent over eight thousand text messages to each other. Colonel Martin was found guilty of fraternization, unauthorized use of equipment, conduct unbecoming of an officer, obstruction of justice, maltreatment of a subordinate, and communicating a threat. Shortly after the conclusion of the investigation, Colonel Martin was transferred away from Fort Polk.

Notice I said "transferred" and not put out of the army or made to retire. I could not understand why he was found guilty and allowed to stay a colonel while I had reported him and was left with a damaged record and the worry that he was still serving. I was told by a senior officer that spring that if I had not worn a wire, it might have been me who was out of the army. As he related it, there were two major factors that saved Colonel Martin from a worse fate. First, he had a powerful three-star general who advocated for him—the same one who wanted the investigation to go away in the first place. The second factor was the army overall was struggling to get volunteers to join, and there had been a series of bad stories recently about leadership. The Army wanted this story to never see the light of a newsroom. That is why today if you google it, you will not find it.

I spent my last few months at Fort Polk working for the garrison commander's emergency services executing an audit of the fire department. It was Brigadier General Bolger's way of trying to take care of me both with a good evaluation and a chance to take a breath. By the time Angela, Hunter, and I drove away from Fort Polk, I had had two straight leaders who were nightmares. But I had also seen the endurance of our young soldiers and the mettle of good senior leaders willing to hold the line. I had learned that surrounding yourself with moral mentors counted and being lucky enough to marry a woman of character were the reasons I was still in the fight.

CHAPTER 16:

IRON SOLDIERS (2009–2011)

꧁ꕥ꧂

Exceptional leaders demonstrate both professional
and personal courage

ANGELA, HUNTER, AND I spent the next year at Maxwell Air Force Base in Montgomery, Alabama, completing Air Command and Staff School. It met the requirement for the schooling I needed to complete to become a lieutenant colonel. Additionally, it exposed me to how sister services thought and to the great community in central Alabama. In the spring of 2009, I was told we were heading back to Germany, and I would be assigned to a lieutenant colonel position as the First Armored Division's provost marshal. Our excitement over returning to Europe was muted by the fact that we knew First AD would be returning to Iraq shortly after we arrived. This meant another year away from my family and a return to a place I had hoped to never see again.

Europe was no longer under the threat of war, so the army in Europe was a shell of what I remembered it in the late 1990s. There were only a few troop units left, and the First Armored Division's last brigade had moved to Texas, leaving a headquarters without troops in Wiesbaden, Germany. The division headquarters was scheduled to depart Germany as well, but first it would go one more time to Iraq. At the time, the army was fully executing a manning system called R4GEN to keep deploying units manned. In essence it meant troops generally arrived at a unit to train up and deploy, and then they would depart after the deployment. This system generated units of almost entirely new people, who then had to create systems and cohesion and share understanding to generate combat effectiveness in a condensed period.

As the First AD staff formed, it was clear that we were extremely young in both age and experience. To complicate matters, our division's mission was unclear at first and began to shift as we prepared for deployment. Upon deployment, the headquarters would swell to well over a thousand troops with hundreds of specified and implied tasks to accomplish. All the primary staff leads were lieutenant colonels except me; I was the only major. This fact would cause me great challenges throughout the deployment, but it forced me to learn to influence above my rank and pick my battles to fight.

My first battle was carving out my role and responsibilities with the division chief of staff. Colonel Mark Calvert was a hard-driving taskmaster who was as smart and detailed as he was merciless and demanding. During my first meeting with him, he disregarded what I told him should be my primary tasks of training Iraqi police, police intelligence operations, and detainee operations as the division provost marshal (senior law enforcement officer). He told me that instead I would serve as the division force protection officer first and do any police tasks second. He further stated he was not sure I should stay a separate staff section and that I would get no additional officers. I left his office feeling like he had kicked me in the face.

On the flip side of my experience with Colonel Calvert, my new senior NCO, Sergeant Major Richards, was exactly what I needed to breathe life into our section. Rich was a hardened veteran of combat in Ramadi and a professional networker, driven to accomplish any task you gave him and fearless in every sense of the word. Together, we cobbled together about a dozen soldiers to form the nucleus of the Provost Marshal Section. We had two more majors besides me, but we lacked midgrade officers and NCOs. I brought on several junior lieutenants while Rich hand selected some of the best NCOs I have ever served with from around Europe. They would serve in positions far above their training and by the end of our tour absolutely thrive at it.

SGM Richards and the author land in Iraq December 2009

After just six quick months of train-up, we deployed to Iraq one day after Christmas 2009. I wrote my second "If I don't come home" letter, cried as I said goodbye to Hunter and Angela, and flicked the switch in my mind to focus on the task at hand. Our division was to take over the security of Baghdad first and then within ninety days take over Al Anbar province from the marines as well. We would be responsible for one-third of the country. We were to close out Operation Iraqi Freedom and begin Operation New Dawn. During this transition, we would place security back into the hands of the Iraqi government.

My section consisted of three lines of effort. First, our police effort would initially manage over fifty military working dog teams, over one hundred contracted law enforcement professional contractors training Iraqis, and all manner of law enforcement support. Second, our detention effort managed the joint interagency detainee investigation / prosecution process and the highly sensitive release program. And finally, our force protection effort conducted weekly inspections across the region and oversaw hundreds of millions of dollars in contracts. The scope and complexity were daunting to me to keep pace. But the team was magnificent. Our NCOs were unmatched and had a tremendous reputation, allowing them to do staff battle with majors and other units better than I could. One amazing example was Sergeant Erica Sides.

Sergeant Sides was Sergeant Major Richards's solution when I said I was going to need a force protection contracts expert. I had no idea what that meant when I said it, but the unit we were replacing said it was critical. I thought it was a joke when this small female MP team leader walked into my office fresh from a line unit. She had no special training in contracts and no significant experience in force protection. However, Sergeant Major Richards's skill was reading people. He told me she was incredibly detail orientated and driven, and once she got her teeth into something, she would not quit. He was right, and Sergeant Sides became a hero within the division, becoming the only sergeant I know to receive a Meritorious Service Medal for the deployment. We estimated she saved the government tens of millions of dollars by correcting bad contracts and holding contractors accountable while reducing wasted troop time. She reminded me that greatness has little to do with rank and everything to do with the desire to do good.

Despite such amazing officers and NCOs within my section, working on a division staff in a combat zone was unlike anything I had experienced before. Each day had all the stress of a war zone without the feeling of connection to the streets that I experienced on my first deployment. However, the pace of work was almost as bad. I was normally at my desk before 0730 and on a good day headed to my CHU (connex living quarters) around 2000 (8 p.m.). After four months in theater, I finally felt confident that I had established a place for the provost marshal and was respected and trusted by the leadership. It was about this time that our commanding general, Terry Wolfe, showed his true depth of understanding of the problems facing the Iraqis. What I didn't know was how it would change my role dramatically.

Major General Terry Wolfe was a tremendous division commander, and I would consider him one of the best leaders I have seen in his ability to lead change through a multitude of leadership techniques and empowerment. As an example, he recognized even before deployment that we would naturally focus on statistics, analysis, and reports about what our forces were doing daily. By its nature a division existed to command and control subordinate units. However, he knew our mission was to get the Iraqi forces to take over 100 percent of the security and support through the country. In order to ensure we understood the focus, he ordered all PowerPoint slide templates in the division to be changed from our motto of "Iron Soldiers" to "It's all about the Iraqis." This meant that on every one of the one hundred thousand slides we used, we were constantly reminded what our focus was. Simple but brilliant.

Now four months into our deployment, as we completed the assumption of Al Anbar and owned one-third of Iraq, he showed his depth of vision and leadership again as we were gathered in a division planning meeting. As all the primary section leaders sat at the table, he began to explain his vision of how the police needed to take the lead across the country from the Iraqi army and how we needed to support them. Of course, as a military policeman, I was thrilled that he recognized this fact, but what impressed me the most was that he had flipped back in his leader's book to pages written before we deployed to lay out his vision. Years later, I had a chance to privately ask him why he envisioned this almost six months before he shared it with us. He said, "We were not ready at the time. We had to build the unit, deploy, assume the mission, and double the mission space." Major General Wolfe taught me that having a brilliant idea is useful only if you know when your team and the environment can handle it.

A few days after that planning meeting, I was leaving the division HQ at about 8 p.m., and Colonel Calvert saw me and yelled, "Why are you taking a half day?"

What I didn't know that night was that the next day he would tell me I was to take over the division's efforts to train and support the police. What I had advocated for eight months prior had now become a reality. I had just got far more than I bargained for. Unlike when I was a battalion S3, I had little context of the tactical situation, and the fact that we had very few US forces capable of validating information left me more than a little unsettled. But the weight and focus of Major General Wolfe drove all the resources in the division toward the police. I became consumed with this aspect of the mission. Luckily, Sergeant Major Sergeant Major Richards and the team continued to hit home runs in our other areas of responsibility.

On a force protection trip to Ramadi, his team discovered that the contract security guarding the infantry brigade's major forward operating base was completely unprepared for an attack. Years of mission creep had totally eroded their capability. Guard towers had no night vision, unserviceable weapons, poor communications, and even dead spaces in their defenses. This was made worse by the fact they were charging the government for equipment that the government was providing. The team's findings led to major improvements and demonstrated to the brigades that this team had real value in protecting the force.

Shortly after this trip, Sergeant Major Richards was conducting an unannounced inspection of the guard force at Camp Victory when he discovered how bad mission creep had gotten. He found four Ugandan contract security guards guarding a pallet of water in the middle of an empty five-acre gravel lot. He walked up to the guards and asked them what they were guarding. They told him they were guarding the water pallet. Sergeant Major Richards asked what they did with the water, to which they responded that they would drink it, and then more water was delivered. It took the team an entire week to discover that two years before, that lot was a staging lot for trucks coming in to be issued to the Iraqi army. However, after the lot was emptied, no one updated the contract security plan, so for almost two years we spent money to guard an empty lot inside a secure base. Sometimes the truth is unbelievable unless you have been to a combat zone.

While our K9 handlers, force protection team, and many of my staff traveled routinely outside the division headquarters, my life involved endless coordination meetings, trips with our generals to see Iraqi police, and seeking medical help for my back. By eight months into the deployment, I was being treated by a physical therapist, getting dry needled and shot with cortisone, and even seeing a massage therapist I found among the contractors. The fatigue was eating my body and mental balance in chunks. Luckily for me, the soldiers I served with there were amazing and family. From field-grade poker nights to lifting weights with my section to cigar smokes with my retired federal agent contractors, we kept one another in the fight.

On one such cigar break with my senior LEP (law enforcement professional), at about 4 p.m. I got an urgent message from the command group. The caller said, "Grab your combat gear, police kit, and meet Brigadier General Mangum at the helicopter pad immediately, nothing else."

I ran to the command group to find out what in the world was going on. Colonel Calvert said there was a standoff between US and Iraqi forces, each blaming the other for killing two civilians. Brigadier General Mangum was flying out to the scene, and he wanted me to go with him to help investigate. As anyone who knows anything about a division provost marshal knows, I had zero special investigative training, and we had zero specialized forensics. I could get the right people, but it would take hours, not minutes. Since I only had five minutes, I asked to take two more personnel with me. Colonel Calvert gave the OK and told me, "Don't screw it up."

I ran back to my office and asked Sergeant Major Richards who we had trained in investigations. Master Sergeant Dave Flynn, aka Grease—because he made everything run better—was accident investigator trained, and my retired ATF agent LEP knew weapons. I told them both to grab their weapons, body armor, and any investigative kit they had. We ran to the helicopter pad just before Brigadier General Mangum arrived. I had not worked with Brigadier General Mangum much before this event. He had arrived late to the division and served as our third deputy commanding general; it was unusual to have three. But before the night was over, I would get to know him well.

If you have never been on a movement with a general officer in a combat zone, you must realize everything happens fast. As a staff officer, you become a strap-hanger and security, and movement is tasked to some unsuspecting ground unit. This trip was no different. We landed at a small combat outpost, where the battalion commander met us and ushered us toward a waiting ground convoy. As he briefed the general, I gleaned there was a US squad and an Iraqi unit adjacent to each other on a busy market street, both conducting checkpoint operations. There had been some type of weapons fire, and in the aftermath, a taxi was found crashed with both the driver and a young female passenger killed.

As we arrived at the snap checkpoint location, Brigadier General Mangum jumped out of the vehicle and went directly over to the Iraqi unit to attempt to speak to the commander. The situation was tense, and in a matter of a few seconds, the general was surrounded by forty or fifty Iraqi military, many of them yelling excitedly. It was impossible to identify the senior commander or even make sense of the situation. I looked around, and the local army unit was all pushed out at the exterior corridor of security, leaving just me and my two folks with Brigadier General Mangum. Before I could even begin to think about attempting to de-escalate the situation, Brigadier General Mangum looked at the senior Iraqi in front of him and said, "Syid"—sir—"I cannot hear myself think, so how about having your folks back the fuck up."

As soon as I heard the message and tone of his voice, I only had a few seconds as the translator began to speak to order the team "About face and off safe."

The three of us formed a tight wedge around the general facing outward just as the translation reached its climax. Instantaneously all fifty Iraqi army soldiers fell silent, and there was a look of shock in their eyes. For about four seconds everything was frozen, and I locked on the nearest soldier wondering if this was going to go south quickly.

A booming Iraqi voice broke the silence and yelled what I later understood as the command for "Attention." The soldiers snapped to attention, and then he ordered them to back up three paces, which they executed smartly.

Then, as calmly and quietly as he could, Brigadier General Mangum said, "Syid, tell me what happened."

We learned a taxi driving by the checkpoint had been fired upon, and both the driver and a teenage schoolgirl in the back seat had both been killed. The US Army Stryker squad said the fire came from the Iraqis, and the Iraqis said it came from the Americans.

Brigadier General Mangum turned and asked me if my team could make sense of the situation. I told him we would do our best. I knew there would be a formal investigation later, and I had to be very careful to not violate army authorities or State Department guidance because Iraqis were now in the lead for their own security and sovereignty. To make matters more complicated, our entire investigative kits consisted of hand sanitizer, a laser pointer, and Leathermans. But it was clear: we needed to de-escalate this tragic situation and begin to solve it before it became a strategic and political nightmare as well.

Hindsight, I am sure we looked silly using laser pointers and wet wipes to "test" weapons for discharge or angles of fire. But at the time, the Iraqis seemed to take us quite seriously and settled down and watched us survey the scene quietly for over an hour. The vehicle and the bodies had been moved, so all we had there were the units and their weapons. Both units had a crew-served weapon mounted on a vehicle capable of hitting passing traffic. If they didn't, their defensive setup would not have made sense, of course. But what was clearly different were the guns themselves. The Iraqis had a 7.62 PK machine gun mounted that was covered in rust, dry, and no sign it had fired a bullet in months. On the other hand, the US unit had a .50-cal remote-controlled weapon system (RWS) that had numerous spent shell casings inside the vehicle (which was normal because US units often test-fired weapons when departing on patrol).

We talked to both Iraqi and US teams manning the crew-served weapons, and both stated they did not fire. The US gunner said he had heard the shots, but he was inside the vehicle manning the remote station. The Iraqi gunner said he had seen the shot, and it came from the US position. However, his claim was questionable based on the angle and distance between the two guns. So I stood up in the hatch of the Stryker, viewed the weapon angles, and noticed the camouflage netting the crew had rigged to provide shade on the top of the vehicle. It was close to the gun as I had the gunner rotate the turret. I asked him if it got in the way, and he said sometimes he had to move it.

We asked to go see the vehicle and the bodies next. The Iraqis struggled to gain the location and permission to see the bodies, which was further complicated by the age and gender of the second victim. But we were taken to look at the vehicle.

It was hard to examine a sedan in body armor, at night with just a flashlight and a Leatherman tool. There were two strikes on the car. One had gone into the engine block, causing massive damage, and we tried in vain to dig it out with the Leatherman. The second hit the window one inch above the dash. That bullet went through the dashboard, hit the driver, went through the seat and into the passenger, then went through the back seat, the trunk, and kept going. After about fifteen minutes of going over the vehicle with no real forensic tools available, we were left with the sense that a 7.62 would struggle to inflict the penetration and damage we were seeing, mainly the engine block.

It was well past midnight by now, and we had pulled back to the unit's FOB while waiting to see the bodies. I told Brigadier General Mangum I didn't need to see the bodies now, and I had some bad news. My LEP, the traffic investigator, and I all saw it the same way— all the evidence pointed to the US accidentally killing the civilians.

Of course, this was not what any leader there wanted to hear, but Brigadier General Mangun and the battalion commander in the room both restated they wanted the truth and transparency. So I explained the three key aspects that had led to our conclusion. The Iraqi weapon had no signs of being fired. The US weapon had been fired recently, and this variant of the RWS still had the "butterfly" trigger on the back, which meant it could be fired without the operator firing it. And the damage to the vehicle appeared to be beyond the scope of a 7.62 round. It was our preliminary assessment that the camo netting rope had got stuck over the butterfly trigger when the gunner was scanning his sector, and when he rotated the weapon back, the rope tightened and depressed the trigger, firing two rounds. It was a tragic accident.

Ultimately the situation was defused that night, and the US took full responsibility. The unit formally apologized to the families and made restitution. Though it was a horrible event, the honor of our army and the character of our senior leaders that night reinforced my thoughts about our professionalism. It would have been easy to cover up, deny, or just not respond to this incident. But that was not a thought that was ever entertained. I had seen my first general officer in a very high-stress tactical combat environment demonstrate remarkable personal and professional courage, and he re-etched in my mind the professionalism our army has and how its values span us all.

This event encapsulated the rest of the deployment. As a unit we worked merciless hours to try to make Iraq a better place while fighting German philosopher Clausewitz's trinity of war—violence, chance, and politics. Being a part of the First Armored Division in Iraq in 2010 was amazing. The mission was massive and the work hard, but the teammates and accomplishment were second to none.

This is about when I started to say, "The army is like crack cocaine—you know it is bad for your mind and body, but when the high wears off, you want another hit." As we landed back in Germany at Christmas, I was proud to be an IRON SOLDIER.

CHAPTER 17:
DEMONS AND ANGELS (2011–2013)

Every person has their amazing story—don't judge the book by its cover.

IN THE SPRING OF 2011, shortly after our return from Iraq, we received the news that I had been selected to command the 519th Military Police Battalion. There would not be much of a break in the pace of our army life. On paper, the division was not doing much but deactivating in Europe and heading for Fort Bliss. But I would have to get the family moved, attend precommand courses, and be ready to take the colors by the fall. However, there was no hesitation for us to accept. I had been given a chance to command—command a unit I had grown up in, command a large line unit near Angela's hometown—and a chance to give back to an organization that meant so much to me. But to do so, I would need the help of angels to fight some demons.

The transition, move, and journey to get to Fort Polk was hard. The army still had troops in two major theaters of war, and units, systems, leaders, and soldiers were extremely tired. I thought at the time that I was handling the continued high pace, stress, and chaos like a champ. Problems arose, I attacked them, crushed them, and moved to the next issue. I was built for this, I thought, and thrived on it. I was a natural crisis leader, or so I thought. However, what I did not recognize was that I was not handling it as well as I thought I was at the time. I didn't notice the signs until my wife, who is always my lead angel, let me know I was not OK.

It was in September 2011, and there was about one month before I took command. I had just crushed another obstacle, something so small that I don't even remember, when she told me very kindly that I needed to seek some help for mental health. By then I had been gutting out the blown disks in my back for over seventeen years, and I was conditioned to just suck it up and push through it. So it was natural for me to just plan on pushing through whatever mental issues she was talking about. But she was not going to let me do it my way. She said, "You have changed after the first deployment, and it has gotten worse after this last one. You get angry quickly over the littlest thing and are ready to kill people." I began to blow her off, and then she said, "Your son watches and learns from you."

Her words hit home. When Hunter was only three, he started mimicking me as I spit my tobacco dip. I had dipped heavily for a decade, but I quit that month on my own. The thought of setting the wrong example was a powerful motivator for me. I already felt guilty for missing almost half his life because of the army so far. Angela was right. Though I never acted out toward her or Hunter, I would not hesitate to explode on something or someone whom I perceived as willfully negligent or failing in a simple task. The target was normally me. If I failed in the smallest detail, I saw it as something that could get someone killed regardless of how trivial the issue. But I was concerned about how it would look as a new commander with mental issues. I was caught between my ego and my anger.

Luckily for me, the army had created a new program called the Military and Family Life Counselors (MFLC). This program sourced licensed mental health experts to specific units and allowed people to see them without adding it to their official medical file. I had just learned of this program when I was going through the precommand and it was just taking off at Fort Polk. I gave them a call and set up a very discreet time and location to meet. The result over the next several months was incredible. Having an objective review of my actions, emotions, and challenges made me a believer in mental health. It was not a magic pill, but it opened my eyes, both individually and as a commander, to how important and powerful it could be. As I started my command, my body was hurting more, but my spirit and joy for the army were running high.

The organization and mission I assumed that fall was massive for a battalion. I had seven organic companies and detachments with over seven hundred soldiers. Additionally, I was ordered to assume command as the director of emergency services for the installation, which added almost two hundred Department of the Army (DA) police, guards, and firefighters to the ranks. And if that was not enough, I was given a third mission of being the task force commander of a rapid response mission in case of natural disaster or attack on American soil, which added two additional companies (three hundred soldiers) that I had to be ready to assume control over in a day. All told I had three different missions under three separate commands and authorities with over 1,200 folks counting on me to not suck. I thought back to Lieutenant Colonel Ted Spain, M. G. Wolfe, and other great leaders I knew. It was here I cemented my belief that climate and culture are everything.

It was a complex time in the army and Military Police Corps. The army was struggling with fatigue and injury from the extended war. I had almost 20 percent of my soldiers in the medical removal board process for various reasons ranging from simple joint injury to massive combat-related damages. The process was extremely slow, complex, and frustrating for both the soldiers and the units. The average time to navigate the process was almost nine months. During this time most soldiers could not do their jobs, and unit leaders had to do double the work to execute the mission and care for soldiers.

Meanwhile, the MP Corps was downsizing its civilian growth from the war and attempting to remerge the law enforcement roles on posts back to MP soldier leadership. This was complicated because over the previous decade, the DA Police had advanced professional standards and expectations and sourced all the key leadership positions with top-notch folks. I would have to overcome personal and professional bias, level the bubbles of training and expectations, and attempt to build a new team without ruining all the good that had come from the last decade. As a leader in a mature organization, you can slightly refine the mission and slowly shift hiring and promotion practices, but what I realized you could affect quickly was climate while providing vision to shape culture.

Luckily for me, the army sent me one of the best NCOs I have ever known to be my battle buddy as the battalion command sergeant major. As a young platoon leader, John Narcisse was my second squad leader for almost two years. We deployed together, trained together, and worked countless law enforcement shifts. He was a man of impeccable character and strong values, a positive role model and an optimist at heart. He was just finishing his first tour as a battalion sergeant major and jumped at the chance to come back to the 519th. When we last served together, I wrote on a Polaroid picture, "Someday you will be a command sergeant major." Now I was overjoyed that he would be my command sergeant major!

When I was younger, I thought climate and culture were some soft nonsense mumbo-jumbo that didn't have a place in the army. As I stared at the units and directorates in front of me, I realized direct leadership was the most useless tool I had to make things better. Instead, I would have to create a vision for what we needed to be, a reason for why we needed to change, and try to encourage the team to shape the path we would take. This meant being as transparent as possible, empowering those below as much as I could and publicly rewarding those who exhibited the traits we were looking for.

This, of course, sounds simple, but it is one of the hardest tasks a leader can do. I was keenly aware of the fact that I had seen hundreds of leaders pontificate on "what they wanted" only to not make a bit of difference after years in command. I drew on the operational planning I had done in Iraq to support the Iraqi police and decided I was going to operationalize my command. For those not versed in operational design, I decided to take a very deliberate planning process to generate the effects I wanted beyond just "doing the mission I was given." When it came to climate and culture, I specified exactly what I was attempting to achieve and then set tasks, missions, and habits designed to reach the effect with leaders responsible throughout. I found it incredibly fulfilling, challenging, and important to the unit, soldiers, and their families. This was when I realized what the next level of leadership looks like. But I made lots of misjudgments while learning.

Shortly after I took command, we began to integrate the civilian police and military police back into one organization. There was a lot of us-versus-them and different standards, roles, and expectations. One day I sneaked into the back of a class where three civilian police were teaching almost sixty new military police on active shooter response. I listened for about ten minutes and began to get aggravated by how both the instructors and students were not focused and taking the training seriously. I thought both were trying to be "cool," puffed-up, and not open to each other. Engaging an active shooter was likely the most dangerous task a responding officer could face. I finally, stood up and said, "Everyone in this room needs to take this seriously, or you will get shot in the face." I visualized a responding officer doing it wrong and dying. I then walked out and went and spoke to my senior leaders about the concerns I had.

We didn't even get a chance to begin to address the issue before I was contacted by the Department of the Army Inspector General Office, who stated they had received an anonymous tip that I had threatened to kill dozens of soldiers and civilians. They were told I said, "If you fuck this up for me, I will shoot you in the face." It was clear someone did not want the merge of the civilian and military police to occur, or they didn't share my same context of what I saw in that training. I didn't know whether to laugh or cry.

The investigation was over before it started because I offered to allow the IG to interview all sixty people in the room to dispel the story. However, this was my first taste of how indirect leadership was different from direct leadership. Most of the people in that room did not work with me daily and did not know my character. All they knew was a ten-second comment from me without context. I recognized it was my job to better message the organization.

I began acting as an echo chamber for my priorities, vision, and goals every chance I had in front of my team. It was an incredibly successful reunification of the police, but my direct leadership style still got me in trouble. On several occasions, I personally stopped soldiers or family members committing offenses on Fort Polk since I was the "sheriff" and had to set the right example. On a few occasions, those I stopped or detained claimed I was exceeding my authority, profiling, or any number of other accusations against my motives. I realized that a lieutenant colonel correcting them evoked a much different response than a sergeant. My boss told me I needed to stop making on-the-spot corrections because it put me at risk, and that was not my job.

I am sure my boss meant his counsel to help me, but I had an absolute visceral response to his words. I almost shouted, *If I am too scared to correct soldiers, how can I ask sergeants or captains to do it? Since when is my career safety more important than army standards?* He had tripped a red line that I didn't understand at the time. I saw not correcting a soldier as a failure in leadership, a failure that could and would get soldiers killed, much like I saw in Iraq. I was determined to not follow his guidance. Every NCO leader of my youth screamed from the past to "hold the line" and lead from the front. But what I failed to understand was that *how* I engaged impacted how I and the institution were perceived by those who didn't work with me daily or know my motives and character. I had to learn to think two steps ahead as a senior leader. I thought back to Mr. Mark Avril and Major Tim Weathersbee for inspiration.

I was always driven by completing tasks, and this meant focusing on data, problems, and plans. Angela, on the other hand, was and is a very intuitive person who has a gift for reading people. When we were newly married, I often discarded her intuition about someone and instead dealt with only the facts I could see, only to be stunned weeks or months later when I finally learned what she knew all along. It was her intuition that led me to select Michelle Ambul as my driver. As I began command, the unit had just assigned a new driver to me, and Michelle was a backup. Angela came by the office one day for lunch and then later told me there was something not right about my new driver.

She got a bad feeling around him and recommended I not use him. She had met Michelle and said she should be my driver instead. My first thought was to argue with her and explain the detailed reasons why the current one had been selected. Well, I had been down this road with my wife dozens of times, and I had learned to listen to her inner voice. I told Command Sergeant Major Narcisse, and he switched up the assignments. It was amazing how many people questioned having a female driver. But not Angela or Command Sergeant Major Narcisse. As it turned out, that first driver, we discovered some months later, had several civilian charges pending against him that he had hidden. In contrast, Sergeant Ambul was an absolute professional. She had played professional basketball in Europe, was methodical, and above all had a strong moral compass.

I found it was important for me over the years to position good people around me with strong positive character and values. Knowledge was not as important as values. I found it motivated me to push myself to not disappoint them and at the same time forced me to evaluate my actions with greater scrutiny. I was so impressed with Michelle that I persuaded her to become an officer, and today she is a major continuing to do great things for the army. I didn't see her true potential when I first met her. This made me think I tried to move too fast to accurately judge people at my new level.

One such misjudgment was our battalion chaplain, Captain Ricky Trull. When he arrived at the battalion shortly after I took command, I saw an overweight, old preacher with a deep southern accent. I thought, *Man, the army must be hurting in chaplains to keep him around.* I told him he needed to get in shape and spend every second he could with soldiers because we had a force that was cracking under the strain of the years, of war and I needed to build hope, energize discipline, and build a team based on trust. As soon as I said it, I was doubting that he could deliver because I had judged this book by its cover.

Over the next six months, Ricky never missed physical training. He lost weight but still struggled with his almost fifty-year-old body in the brutal Louisiana heat. But he never complained. He just got after it. Additionally, I noticed every time I had a concern about a soldier or family, I found out Captain Trull was already engaged. He was never in his office; he was instead out with the units almost every day. He was quietly doing what I had asked him to do without making a big show of doing it. He had even impressed Command Sergeant Major Narcisse, who told me, "That chaplain is a combat multiplier." I didn't understand his motives, but I was determined to recognize and thank him, so I asked him to come by my office for counseling.

Later that week he reported as directed. I told him I was super impressed with his first six months in the unit, and I just wanted to tell him. He said thank you, and he was happy to be with soldiers. I asked him why he was so focused and driven after apparently entering the army so late in life. He told me he had been a small-church pastor and that his son became a soldier. I had not known that until this point. He said his son fell into depression under the strain of everything we were facing during the war and committed suicide. It was then that he decided to join the army and make sure that no other soldier he met would feel hopeless. I was stunned and humbled at both the faith and humility of this man of God.

This world is full of demons. Demons take shape as terrorists, fear of failure, regrets, unrealistic expectations, and a thousand other ghosts of doubt. Over my two years in battalion command, I had the most fun I had ever had in the army while being so exhausted. I felt like I was a small part of a team beating back those demons and making life better. As I look back now, I realize it was the angels God placed in and around that unit like Ricky that made all the difference. I learned a senior leader's job was to create a good climate and try to positively affect the culture, hire good people, and then take care of them. It was a simple formula.

The Viper Battalion reminded me there is no such person as "The Army"; instead, the army is made up of a million unique and amazing individuals. A leader's job is to serve them. I am super proud to have been called Viper 6.

VIPER Battalion and DES key leaders pose for aerial photo

CHAPTER 18:

DANTE'S NINE CIRCLES OF HELL (2015–2016)

❧

*Every person must recognize what they
are simply not built for*

AFTER I SURRENDERED THE colors of the 519th MP BN in the fall of 2013, I was, fortunately, selected for war college and moonlighted as the deputy chief of staff of Fort Polk until I departed. As a family, we selected to go to Air War College in Alabama because when we attended ILE there, we enjoyed the area. Air War College was a fantastic break from army life. Arguably my year there studying strategy and strategic leadership was the easiest year I experienced to date. However, My air force vacation would slam back into army reality when the army tagged me for a "black book" assignment in the Pentagon.

In the spring of 2015, we packed up and headed off to Washington, DC. I was assigned to be the deputy to the administrative assistant to the secretary of the army's executive officer (XO). As a three-star equivalent senior executive, the position rated a lieutenant colonel

XO. One- and two-star generals and senior executives were given aides, but at the three- and four-star level, they were called XOs. So I became an aide or personal assistant at forty-two years old.

My principal, Mr. Mark Averill, was one of the most ethical, dedicated, and focused leaders I have ever met. He had forgotten more about the army than most generals would ever learn. He was in his office before the sun rose, and he routinely didn't head for home until seven at night. Each time I got to sit in the room with him, it was a free education in the required bureaucracy and enterprise level of the army. Looking back, he was an amazing role model and professional, but I disliked almost every aspect of working at the Pentagon.

If you have never been near the Pentagon, the idea of working there might sound cool. However, I don't think I ever met many who worked there who would share your perception. In 2015 there were well over twenty-five thousand people who worked or traveled to the Pentagon daily. The basics like traveling to work, parking, and eating all required careful and deliberate planning to avoid pain. As an example, I lived at Fort Belvoir, eighteen miles away. If I left for the office at 0500 in the morning, it took twenty-five minutes to drive there. If I left at 0520, it took forty-five minutes, and if I left after 0530, it could take anywhere from one to two hours. We tended to look at any travel off Fort Belvoir like a combat patrol, leaving hours early for any event and always prepared to sleep overnight.

I had a great boss. The area we lived in wasn't our cup of tea, but what was a deal breaker for me was the sheer bureaucracy and waste I saw every day and feeling like we were spinning our wheels. This is not an attack on the senior leaders I met; it is the pure effect of many great leaders following the rules and policies laid out by our government. As any student of bureaucracy knows, it never shrinks and focuses on the process versus product. This is a necessary construct for any large organization, but personally it made me mad, and I longed to get back to troops and the tactical level.

There are two stories that highlight the insanity of the Pentagon that I often reflect upon.

The first story occurred when my boss assumed control of the agency that handled computers and automation for the army within the Pentagon. This was a massive agency, mainly contractors, that was morphing into a joint agency over the next year, and the army was trying to understand the equities, roles, responsibilities, and challenges within this agency. The senior leadership of the agency dissolved, and only a few GS15s were left as the transition began.

My boss pored over wire diagrams, tables, charts, and manning reports in a furious quest for knowledge. He met with dozens of employees, leaders, contractors, and clients to help shape his understanding of the mission set. I admired his breadth of inquiry and depth of research. If he had not been so focused, I doubt this story would have ever existed. But I clearly remember the day that he was trying to understand the roughly forty-eight government employees that were remaining in the agency headquarters until transition. He had not been able to understand the role of a five-person section called something like "analysis and assessment." He finally just directed the lead of that section to come see him, and I happened to be there.

The department lead came in, a lean, middle-aged man in a coat and tie, and sat nervously at the end of the small conference table in Mr. Averill's office. After several rounds of social banter and mild discussions about general work topics, Mr. Averill pointedly asked what the department did. The lead manager—I will call him Mr. Smith—said they were responsible for producing the report evaluating the contract company's performance. This sounded very reasonable and important to ensure the multimillion-dollar contract was executed correctly. I would have been satisfied with that answer and moved on, but I guess that is why I was not in charge.

Mr. Averill asked how a five-person shop evaluated thousands of service calls, hundreds of contractors, and millions in expenses. He had noticed that the department did not have analytics experts or statisticians, and it appeared to be a tall order for such a small team. Mr. Smith responded that they did not do any of that. Mr. Averill locked his gaze onto Mr. Smith and asked him to explain. Mr. Smith said their job was to produce the report for the rest of the agency, not to execute the analysis. This clearly puzzled Mr. Averill, and he further asked who did the actual analysis. Mr. Smith responded softly that the contractor did that.

In a nanosecond the environment in the room shifted. The silence was so deafening I thought I had heard him wrong. Mr. Averill's entire face went red as if the temperature had just increased five degrees, and the man began to sweat and avoid all eye contact. Mr. Averill fought not to share his emotions with Mr. Smith and did a great job. But after months of working directly with him, I could tell how upset he was at this information.

Over the next ten minutes, Mr. Averill grilled him about who came up with the rating criteria, who collected the data, who analyzed it, etc. As he finally let Mr. Smith depart, Mr. Averill turned to me and said, "Simply amazing. We have five highly paid government employees whose only job is to forward a report produced by a contractor. The report was on that contractor's performance based on metrics designed, collected, and evaluated by the same contractor. It is no wonder they have perfect scores."

If this story seems shocking, then this next story involving the Equal Employment Opportunity (EEO) Office will seem like pure fiction. The EEO office was another department that fell under my boss's purview. I dealt with them regularly because they were always late on actions, full of errors, and the least informed team. Almost weekly I had to delve down into the department to get routine and mundane tasks accomplished. Since the thirteen-person department was staffed from GS 12 to 15 (all middle and senior levels), I assumed the section was just overworked. Every month they were late submitting the EEO monthly poster design for approval.

For almost four months I thought this office created the monthly EEO posters I saw across the army for Black history, Asian Pacific history, women's history, etc. I was amazed that we seemed to approve these super late, but somehow no one ever complained about not getting their posters. It wasn't until my fifth month there that I learned that up to four people in our EEO office spent all month designing a poster that just went on one pillar in one hall of the Pentagon. I was flabbergasted and asked why we didn't use the ones that went everywhere else in the army, to which I got a blank stare back from the department chair.

After I discovered this blatant waste of energy, I began to try to understand why this department appeared so dysfunctional and overworked. Mr. Averill had much bigger issues to work on but allowed me to investigate the department in my spare time. I learned the department did not make enterprise policy but instead acted as any EEO office would at an installation, focusing on handling investigations and claims of discrimination. The first thing I discovered odd was that the department was working cases from around the world. When I asked, they said if someone called them, they would work the case. I asked them why they did not refer the case to the geographic offices around the world, and they again looked dumbfounded. I realized the personnel, though senior in pay, had almost no understanding of the construct of the army.

Based on this revelation, I pressed them for data on the cases they had worked on over the last twelve months. After several weeks of prodding, their deputy sent me a spreadsheet of all their cases. After I confirmed that it was complete, I only needed five minutes to analyze it. A department of thirteen people had conducted fewer than fifty investigations over twelve months. Of those fifty cases, almost twenty were outside their purview and should have been transferred. But the fact that they only worked on thirty cases inside HQDA was not the most shocking piece of info. Who they investigated was.

For months I had burned hours and hours propping up this section far beyond the scope of my duty position. Countless times I had to redo their work or help them with the most basic procedural task. And now, in front of me, I knew why. Our HQDA EEO office worked on over a dozen cases filed by the EEO office employees themselves, many against one another. I just could not believe it. HQDA was made up of over five thousand people, and our EEO office of thirteen people only worked fewer than twenty cases for them in a year, or .4 percent, while executing twelve cases from their own thirteen-person office, or 92 percent. It was literally the self-licking ice cream cone. Here we had an agency designed to help others using most of its time investigating itself.

If soldiers had been caught doing these actions, their punishment and admonishment would have been swift. But, inside the Pentagon, I learned that even the best SESs and GOs were almost powerless to influence this behavior. The system that was designed to provide continuity through a large civilian workforce had translated to a system where it was impossible to hold them accountable for unprofessional conduct.

I cemented two ideas in my head. First, the idea that it was cheaper to have a civilian do a task than a soldier was false. Young E5s could have easily executed the tasks these senior civilians were doing. Second, I would never come back to the Pentagon regardless of the cost.

As I was discovering these life lessons, the army gave me more to chew on. I was selected for colonel, but only an alternate for brigade command. This news meant I was no longer considered a top performer. To make matters worse, my branch contacted me and told me they needed me to deploy for a year to Afghanistan. I had not deployed in over four years, so I knew I was at risk, but I also knew almost a dozen colonels in my branch who had not deployed in much longer. That was when I realized those colonels could retire as colonels. I could not. You must do three years as a colonel to get full colonel retirement pay. This made the math easy for our family—retire with lieutenant colonel pay, or deploy for another year for a colonel's retirement. I jokingly said I volunteered for Afghanistan to escape the nine circles of hell.

CHAPTER 19:
ALICE IN WONDERLAND
(2016–2017)

※

*War is politics, and politics often don't
make sense to a warrior.*

I HAD MIXED EMOTIONS as I prepared to deploy to Afghanistan. On one hand, I was extremely tired and becoming bitter toward the army. I felt like I had done more than my share and wanted to find one of those jobs that I always heard about where you could go home before dark. Additionally, I dreaded another year away from my family. If you have never spent an entire year away from your immediate family, it is hard to describe the guilt, loneliness, and isolation you feel. However, I had never been to Afghanistan, and I had a burning desire to be a part of that theater, much like I had been in Iraq.

I was going to be part of a NATO advisory team made up of about a dozen lieutenant colonels and colonels from six different countries who advised key Afghan politicians and generals on the strategic level at the Ministry of Interior. I would deploy as an individual and meet my teammates once I landed in Kabul. I was assigned to be the personal adviser to the deputy minister of security, an Afghan four-star general who commanded 156,000 police. I received no training on the culture or people of Afghanistan, no course on how to advise, and knew nothing about NATO—what could go wrong?

In late May 2016, I flew out to El Paso, Texas, to process through the generic deployment center. After a week of training and screening designed to frustrate any soldier who had ever deployed before, I finally was directed to board a plane heading to Kuwait. After playing the duffel bag shuffle multiple times, dragging hundreds of pounds of gear from place to place and plane to plane, I finally touched down in Kabul, Afghanistan. I was struck by the rugged beauty of the landscape and the thinness of the air at the same time.

NATO at large in Afghanistan and my small advising team were both filled with strange functions, alignments, and efforts generated by the complexities of war, bureaucracy, and a massive mission set. We could observe air strikes via satellite and then go eat soft-serve ice cream. We could FaceTime home and hit the gym but couldn't go five yards beyond the headquarters without expecting to get attacked. There were cigar nights and embassy parties mixed with tactical combat patrolling and a lot of counterterrorism direct action. After more than a decade of war, I felt like Alice arriving in Wonderland. It was like nothing I had ever seen before or expected to see. Politics, policy, bureaucracy, parochialism, and Clausewitz's holy trinity of war all sat at the same table.

I was assigned to advise General Abdul Rahman Rahman. He was the number-two leader in the Ministry of the Interior, second only to the minister himself, who was advised by then Colonel Duane Miller. General Rahman was a massive man rumored to be a prolific judo and wrestling practitioner. The first time I met him, I was struck by his massively long arms and large hands. Additionally, I noticed his face portrayed his emotions through large gestures. He stood in contrast to the minister himself, who was a lean and small man. This stark contrast, along with their personal characters, led us to call General Rahman Donkey Kong (DK) and the minister Luigi, from the famous Donkey Kong video game.

I could write an entire book on my experience with General Rahman and the ministry. The closest analogy I can make to the experience is that it was like having a completely out-of-touch consulting firm working with the mafia. NATO was the out-of-touch firm, and the Ministry of the Interior was the mafia. And of course, I dealt with one of the mafia bosses every single day. Both organizations were full of wickedly smart people who had to play the cards they were dealt. NATO unfortunately had a losing hand from the beginning when it came to winning against both the mafia and the Taliban.

Of course it is easy to poke at NATO and its inability to produce success in 2016, but when you do the math, it is really a testament to the individual leaders there that anything was accomplished at all. First, coalition forces were no longer on the battlefield or "holding terrain." This meant the fight was solely executed by the Afghans, and there was no independent way to verify Afghan government reporting on thousands of projects, engagements, policies, and missions. This was very similar to what I had seen in Iraq in 2011, but the level of violence and enemy action was tenfold in Iraq at the same period of their evolution to self-security.

Second, NATO member policies, or caveats as we called them, further hampered what few troops we had available to gain context of the operational environment. Some countries would not travel outside NATO base camps, some could not travel by ground, some were limited in the types of missions they could do. I would venture to say there were only five or fewer countries present whose soldiers had the flexibility to be relevant to the changing conflict. This meant any NATO team built operated on about 60 percent strength in any set mission. I couldn't imagine a warfighting headquarters designed this way. However, I was reminded by a general that the fact that NATO was together in Afghanistan and functioning was a victory. It was a strategic success, but an operational nightmare.

The third factor completing the chaos of NATO was the bloated mission creep of good ideas. After a decade of expected progress, more and more nation building and national pet projects consumed the bandwidth of leaders. As the Taliban overran police districts and the Afghan government struggled to provide water, food, and ammo to its forces, we had meeting after meeting about women's rights offices in police stations, school construction, and drug addict rehabilitation. Hundreds of millions of dollars had been spent by various NATO countries on these projects, and their people wanted value for their contributions. These ventures were started based on the assumption that security would hold or increase, but starting in 2015 and continuing in 2016, it was clear that maintaining basic security was becoming a bridge too far.

For my part, I jumped right into the deep end of the pool trying to satisfy NATO expectations while quickly realizing the futility of most of these projects. I remember being grilled by a British embassy liaison on women's rights when I came back from a visit to a local district police headquarters outside the capital regarding the status of the sexual assault office. She wanted to know how it was staffed—did they have their own police car, computer, private exam room, etc.? She could not believe me when I stated the police station itself had no power, no internet, and barely enough police to protect the headquarters, and I had seen no sign of any female police. I saw it went counter to the narrative she had believed and passed along to the embassy for years.

The British two-star general over my team was an eternal optimist and acutely politically focused, while most of my peer advisers from the US, Italy, Australia, and Romania were seasoned leaders with decades of experience who quickly became skeptics as to the success of the programs and projects they inherited from the previous teams. We struggled to understand the mission, the country, the Afghan government, and the detailed and complicated linkages. Meanwhile, the "mafia" we advised had dealt with dozens of rotations of NATO members and were experts on knowing how to play us based on experience, national desires, and the time of the year.

During my first ninety days in the country, DK was exceptionally difficult for me to advise. He played his cards very close, did not let me into his inner circle, and rarely asked for advice or shared his thoughts. At the same time, I was personally hurting. Beyond the normal depression, isolation, and stress of any deployment, my chronic upper-back pain had me thinking I could not finish the deployment. The ruptured disc in my neck, the daily wearing of body armor weighing almost fifty pounds, and computer work left me in tears constantly. I was visiting the medical clinic two times a week for needling and pain treatment while eating Motrin and arthritis medicine like candy. I told Duane one night in August that I wasn't sure I could keep going. However, I could not bear the thought of tapping out and not finishing the mission, so I did what I had to, to stay in the fight.

But enough about my pain. I think the best way to explain the insanity of Afghanistan is to tell the story of just one mission. In November 2016, Afghanistan and Turkmenistan's joint railroad venture generated a ribbon-cutting ceremony, which both presidents would attend. The deputy minister of defense and security had the responsibility of ensuring the ceremony was protected and secure. Both generals were to travel to the border to personally oversee the security. Both the deputy defense minister's American colonel adviser and I were to travel with them to Mazar-i-Sharif in northwest Afghanistan.

Here it is important to understand that the special authorities we had to travel were unlike every other NATO adviser. When any other adviser traveled to advise Afghan forces, they had to be transported by a coalition patrol of a dozen soldiers and once there had a personally assigned guardian angel (bodyguard) who went everywhere with them to protect against insider threat or kidnappings. However, there were four US advisers, including me, who had waivers signed by the NATO commander that allowed us to move without any NATO security. This meant we would travel with just Afghan forces. I did this several times a week in and around Kabul, but when we traveled outside Kabul, it was far more intense. I carried a cell phone, satellite phone, beacon, rifle, and pistol knowing that no NATO force could respond in under an hour if I got in trouble. I didn't speak the language, which meant I never stood a chance of identifying rising tension or threat before it would occur. This would have been exhilarating when I was a lieutenant, but it was just unnerving as a colonel.

Our small team, consisting of the two generals, their personal aides, and the two of us, boarded a Russian-made Hip helicopter flown by Afghan pilots at first light. As the helicopter took off, the temperature plummeted, and soon we were stretching the ability of the helicopter as the rotors fought for lift in the thin air over the snow-covered mountains of the Hindu Kush. Unlike the well-maintained US aircraft I was used to, this helicopter looked like it was on its last leg except for the ridiculous Afghan rug adorning the floor. I faked looking calm until we finally touched down a few hours later.

Mazar-i-Sharif was cold. I mean a cold that I didn't know could exist. The army corps headquarters we landed at was covered in three inches of ice, and the wind was blowing 20 or 30 mph. I understood why Afghans traditionally didn't fight in the winter—obviously they were too busy trying to just survive. The corps headquarters itself lacked heat except for the commander's office. A headquarters that commanded tens of thousands of troops could not or would not heat their own HQ.

Our generals traveled to the border the next day after a late night of meetings, but we had to remain at the corps headquarters because the border was beyond coalition air cover. The corps sergeant major was charged with our security. We slept in the commander's office, and four personally selected soldiers were our protection. The corps sergeant major was an impressive man with a massive handlebar mustache who was educated and trained by the US. To pass the time, he took us on an amazingly surreal trip.

The Germans oversaw this portion of Afghanistan but no longer traveled to advise Afghans. So most local Afghans had not seen NATO forces in over a year, and there had been no verifiable threat analysis done lately. However, we knew this region was fiercely anti-Taliban, and the sergeant major was well connected in the region. So when he suggested we go see the famous Mazar-i-Sharif prison where CIA agent Johnny "Mike" Spann became the first American killed in combat in Afghanistan in November 2001, we jumped at the chance.

So, as the sun crept over the mountains, the three of us hopped in an unarmored Toyota 4Runner and set off. It was a weird sensation between believing it was incredibly dangerous to travel there in such a manner and feeling that we were as safe as we could possibly be in Afghanistan. These emotions clashed as training, common sense, and my gut all tried to voice their opinion. In the end common sense won out, and the sergeant major was both vigilant in our safety and an excellent guide to this now historical site.

The prison itself stood as a symbol of what this war had become. There was a small monument to Agent Spann in the middle made of beautiful marble that was now worn, dirty, and decaying. As we walked through the large courtyard where the massive battle took place, everywhere we looked, there were piles of human feces, spent shell casings, and human bones. It was a graveyard in every sense of the word. The videos you can watch online of the battle do not do justice to the scale of the battlefield. The Taliban were surrounded by forces on the high ground with just a few hundred yards of virtually open terrain to defend from. Yet the fight went on for days, often with just a few meters between combatants, until the final fighters were ferreted out of a basement. I was deeply moved by being there.

That evening my thoughts of the cost of this war were reinforced when our generals returned from the ceremony. There was not the usual euphoria that should have accompanied a successful event such as the one executed that day. We quickly found out why. A helicopter from the unit that had flown us the day prior had crashed after liftoff, killing a corps commander and friend of our generals. I think everyone was reflecting on such a senseless loss, that it could have been us, and now, since all helicopters were grounded, how would we get back to Kabul? There was nothing NATO HQs could do to help in the short run, and I realized we were along for the ride wherever it was going.

The Afghan solution was to ground move to the Mazar-i-Sharif civilian airport and catch a civilian chartered flight back to Kabul. The president's personal protection battalion had a chartered flight heading back to Kabul that evening, and the deputy minister was making arrangements to get us all on the flight. As we arrived at the airport, it was obvious we were the only two foreigners seen in a very long time. We were surrounded by hundreds of armed Afghan forces from dozens of varying units. Even though it was a civilian airport and the charter was a civilian plane, it felt more like a military base because everyone was armed.

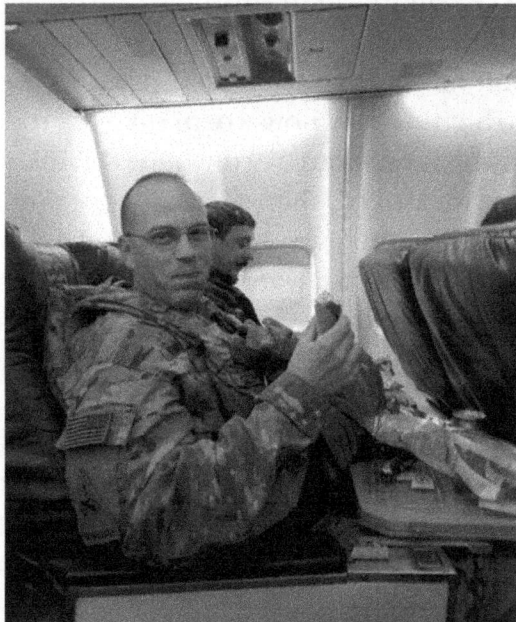

Author in business class with fully armed and wearing body armor

It was dark when we were called to board. Soldiers were running toward the plane like it was the last flight out, and our small party was pushed through the sea of troops to board. We were sat in first class at the front by a beautiful stewardess. It was like something out of a war comedy. She went over the emergency procedures on the speaker, served us a preflight soda, and then politely asked me if I would take the magazine out of my rifle. So there I sat, in full body armor with my rifle across my lap (unloaded) and 9mm still locked and loaded on a civilian aircraft. I thought the TSA was going to jump out at any moment. It doesn't get any more *Alice in Wonderland* than that.

When we landed in Kabul just an hour later, I was faced with yet another significant problem. We were on the civilian side of the airport. My base was on the military side, and of course I couldn't just walk across the runway. It was past midnight, and I had three options. I could go with an Afghan colonel I had never met, who I was told could get me across to the military side. I could go back to the ministry and wait five or six hours for NATO to spin up a patrol to come get me. Or I could go back to the ministry, and the general's aide, whom I trusted my life to daily, would get me back to my base camp. I chose option three, as at the time it seemed the best. Meanwhile, I was texting Duane the plan so that if I vanished, NATO would have a starting point to find me.

The ministry was located on the west side of the Kabul airport and maybe four miles from the NATO base camp known as HKIA. The route between the two was a limited access route, which meant that it was much safer than the rest of Kabul. However, in the middle of the night, it felt like I was on the dark side of the moon. It didn't help my nerves that the aide pulled up in a little white sedan with a young driver carrying an AK47 that I had never met. However, I had already put my life in his hands a dozen times over, so I slid into the passenger seat and said a small prayer.

We could not drive up to the main gate guarded by the Turkish military. A white sedan at that time of night would at worst have been fired upon or at best subject to detention for hours as they tried to figure how a US colonel was driving up in an unmarked, unarmored civilian car with no escort. So we were going to enter the Afghan airbase and then cross from the Afghan side into the NATO side since the guards there were Western contractors and US soldiers. It sounded like a great plan. Unfortunately, the Afghan air force guards did not get the memo.

As we eased up to the airbase gate, we were met with the expected bright lights, rifle barrels, and challenges. The first guard approached me yelling something, and my Afghan partners yelled something back. Not speaking the language meant all I understood was the pounding of adrenaline in my ears as the guard circled behind me continuing to speak aggressively. As the aide told me they did not believe I was an American colonel and thought it might be a trick, I slowly moved my selector switch to semi from safe on my M4 and twisted to the right as far as I could in the seat. It was meant to make me feel a little better, but I knew it would be meaningless if it went kinetic—I would be dead before I could react.

The aide jumped out of the car and began to de-escalate the tension by showing his ID and explaining who he worked for. The air force guard went back to the gate shack and began talking on the radio. So we sat there for a little over ten minutes while the guard waited for guidance from someone above. Finally, we were granted permission to enter the base, and we made the short drive to the gate dividing the Afghan and NATO sides. There again we were met by spotlights and muzzles as the guards reacted to such an unusual appearance of a car at this time.

I told the team to not approach the gate, because I was not sure of the procedures there and did not want to be shot by accident. I thanked them for getting me home, and I hopped out of the car and walked the last seventy-five yards toward the gate. The contractors and young troopers looked absolutely stunned. They began to ask me who I was, what was my movement number, and unit. I pulled out my "magic" waiver card I kept on my ID tags and said, "It is a long story that I cannot tell you." I must have looked like some super covert operative to them, but I felt like a scared, tired old man in the middle of an adrenaline dump.

As I finally texted Duane that I was safe, I took stock of what I had just done, why I had done it, and whether it would make a difference in my mission, NATO's success, or Afghanistan as a whole. This was my Afghanistan. A strange trip into Wonderland, where we did what we were told to do the best way we knew how but knowing deep in our consciousness that the beauty of the local people and terrain could not overcome the beast of corruption and tribalism. As the winter darkened the skies, so did my mood. I had lost friends here, my DNA pushed me to never quit, but the outcome felt inevitable.

As we ushered in the new year, I was looking forward to finding one last assignment far away from DC where our family could settle. I knew I had to do two more years to qualify for O6 retirement, but one thing I was sure of was that I was not going to do it one day longer. My body hurt, I was depressed, and I was risking my life daily helping the Afghan mafia scam America. I was not in a good place, but thankfully, like so many times before, the soldiers to my left and right made the difference. There is something bonding about shared misery. That's why they were the first people I told when I got a call that changed everything.

CHAPTER 20:
EV 4 LIFE
(2017–2019)

❦

Good climate and culture are everything, and
Good leaders take care of people who help create it.

IN LATE JANUARY, I got a weird call from an old colonel in the Provost Marshal General Office asking how my back was and if I was deployable. I laughed as I cussed him and told him I was deployed wearing body armor every day. After several more unusual calls, I finally learned the army was looking to put me in command again. I told them no, there was only one unit I would have considered commanding, but it was deactivating. That is when they told me it was not and that the Eighteenth MP Brigade, known as the Ever Vigilant (EV) Brigade, needed a commander that summer. We had a chance to go back to Europe as a family and command the most decorated MP brigade in the army. We said yes.

The Eighteenth MP Brigade was the unit that took back the US embassy during the Tet Offensive in Vietnam, it was the unit that commanded all military police during Iraqi Freedom, it was the unit that my mentor, Ted Spain, had commanded. It was a dream to command it, but first we had to jump through army hoops to get there. It was late March before I got orders, and I had to get out of Afghanistan, attend precommand course, pack up the family, and move from DC to Germany in less than seventy days. Our family reunion consisted of packing boxes, updating passports, and studying for the European driver's license test.

The first sixty days of command were a whirlwind. The unit was broken and stripped down to 30 percent staff, and it was suffering from several years of atrophy. It had been years since the unit had done anything much beyond garrison policing, and it lacked tactical equipment, focus, and most of all, a warrior spirit. Making things more challenging was the complexity of the organizational alignment. I was both the USARUER provost marshal and Eighteenth MP Brigade commander at the same time. On the one hand, I was the senior army police officer in Europe setting policy and executing operations daily, including overseeing army law enforcement, contract guards throughout Europe, and the driver's licenses and plates for all forces. On the other hand, I had the largest MP battalion in the world, the only construction engineer battalion in Europe, and the Department of Defense Regional Prison. I had over three thousand folks spread across fourteen installations in seven countries counting on me not to suck.

The funny thing about commanding at this level is that it is where you have the least amount of power, and at the same time, it is the place you can get to do the best. This may seem odd to many reading this, but at the O6 level, everything you need to help your unit is owned by another O6, so you must ask, negotiate, and work well with others. However, if you do it well, the results are amazing. It was the first time that I didn't have to ask my boss for permission. Major General Shapiro, my direct supervisor, who was a logistician, once told me, "I am not smart enough to tell you what you need to do; however, I am smart enough to advocate for what you tell me you need to do." Between him and then Lieutenant General Hodges and Cavoli, I was set up with a fantastic environment and a chance to make a difference.

I can summarize the most important lessons learned in my two years in command in one simple statement: people matter, and culture and climate are everything. Now, I will be the first to admit this was not the same understanding I had when I was young in the army. Then it was the mission above everything. Now, as I was older, I realized it was not my job to complete the mission, even if I was responsible for its execution. My task was to set the conditions and resources to allow great folks to do great things. I felt that I was built for this assignment. It was by far the most fun I have ever had in my career.

In two years, we rebuilt the tactical fighting capability, and the BDE spent more nights in the field than the entire ten years before. We radically shifted the way we did law enforcement and construction and how we prepared to fight at a moment's notice. The global war on terror was winding down, and focusing on the growing Russian threat was creating energy. I had been blessed with an almost blank slate—I got to restaff a new team, focus on a new threat, and design how we would contribute to the greater army.

We created our vision and designed an operational approach that had both culture and a warrior mentality woven into it. Then we vigorously messaged it, empowered as many as possible, and rewarded often. Setting a positive climate while trying to understand and nest a plethora of cultures both internal and external is an all-encompassing task. I learned there is a vast difference between a leader's aspiration and doing the work to make it happen. I think that is why most tend to focus on the mission instead—because it is much easier to attack and measure success. However, they miss the point that great people with focus will complete the mission, but a complete mission will not guarantee great people with focus.

I could write an entire book on just the amazing folks I served beside in the brigade. My command sergeant major, Ted Pearson, was a warrior philosopher. My battalion commanders and sergeant majors were stronger, smarter, and faster than I ever was. My brigade staff were young, dedicated, and people of true integrity and honor. And the young company commanders and first sergeants were absolute heroes daily. I've always had a little ADHD and OCD, meaning I've struggled to exercise any patience and always figured I could take on one more task, which made me the weakest link. No one knew this more than my driver, Sergeant Stras. To know Sergeant Stras is to understand the soul of our army and the caliber of soldiers who make it the greatest force for good in the world.

As I took command, then Specialist Stras was recently assigned to be a command driver. Our drivers played a huge role in running the office, driving the command team and planning our travel. In my case I had to travel almost every week, many times to multiple countries and a huge variety of events. At the same time, I needed to read stacks of documents, take meetings, solve problems, and continue routine operations regardless of whether I was in a snow-storm traveling through the Alps or flying back from visiting troops in Bosnia. To just keep up, I had to use every single second I had on the move.

Specialist Stras was promoted to sergeant shortly after my arrival to the unit and became my primary driver. By our second month together, we had logged over two thousand miles in a Humvee and twice that distance in our government van. We executed thirty-hour straight movements in tactical exercises and crisscrossed almost a dozen countries. I would describe her ability to my wife as a "fire and forget" missile. She asked the right questions, checked her facts, planned the small details, was never late, and was always prepared. I learned she came from a first-generation American family and that most of her family still lived in Poland. She spoke multiple languages and could effectively understand several more. She once was married and worked for a prominent brokerage firm in New York. When she finally decided to divorce her husband, she wanted a total change from the corporate world, so she enlisted in the army as a military police officer.

But as with most things in the army, the strongest bonds are often formed in the most chaotic environments, and this was true for Stras and me. I had not been in command for more than a week when we had to depart for a major NATO exercise. We were driving a portion of the Brigade eight hundred miles in tactical convoys, assuming command of additional forces from seven different countries once we arrived in Romania, and then executing an opposed river crossing at night. There was no more challenging and dangerous mission an MP unit could attempt. The only catch was the brigade was untrained, and the previous command team had done absolutely nothing to prepare.

As we departed Germany, my command vehicle lacked any way to effectively communicate beyond FM, making me a spectator as opposed to a commander of anything. Our tactical trucks were stuffed to the legal limits with everything we would need, as we were pushing the range any unit should attempt to move in one action. I tried to settle my mind but found it difficult because only a few months ago, I was doing this in Afghanistan, and now I was in a totally different environment. As I tried to think ahead in the operation, I was snapped back to reality by the bang right next to my head. Sergeant Stras had veered a little too far to the right and hit an on-ramp sign with our side mirror, sending it into a thousand pieces. She apologized profusely, and I thought, *If she only knew how many mirrors I had killed.*

After three painful days getting beat to death on the road for fourteen hours a day, we finally arrived at Valcea, Romania. We met all our NATO allies and partners as they arrived throughout the next twenty-four hours and began attempting to deconflict communications, logistics, ammunition, and a million other issues. The Macedonian Humvees took a different type of fuel, the Slovenians' blank adapters didn't fit, the Romanian radios would not connect to ours, and so on. At the same time, I discovered the exercise had been designed with almost no oversight or construct. On the one hand, I could do just about anything, but on the other hand, I had no idea as to the capability of my forces or their limitations.

On the positive side, our allies and partners had sent their A-teams. Of special note was the young but massive Romanian lieutenant who served as my liaison officer to our host country, Bila. He was energetic, focused, and absolutely passionate about completing his mission, and his mission was to make our training a success. His energy was matched by the Romanian and Polish infantry as well as the skill of the multiple bridging units from four different countries. However dedicated and professional these forces were, they could not make up for the absolutely crazy training event we were going to attempt.

The loosely designed mission parameters set up a scenario where an enemy had destroyed all the bridges in the region and had inserted blocking units to keep the routes closed, so NATO could not reinforce frontline units being hard pressed. Our brigade was ordered to secure a crossing point and build four temporary bridges across a river over one hundred meters wide to allow a Romanian brigade heading our way to cross and push toward the front. This scenario was one of the most difficult tactical tasks you could do but had never been done with an MP BDE leading it. Let alone an MP BDE with foreign units that it had never met or trained with before that week.

To make this mission more complicated, we were going to do it at night. At night in a civilian town with over fifty thousand people in an area that was not sectioned off from civilian traffic. We were going to try to fight across the river and then move into the bridgehead with hundreds of bridging assets to build not one but four bridges while dodging civilians, protecting property, and fighting a mechanized force on the other side. The Second Armored Cavalry Regiment (2CR) unit had attempted the mission just two days before us and failed to ever get the bridges in place. A pure US force of several thousand troops who trained together daily for over a year had failed, I knew we had to approach it differently if we were to stand a chance.

2CR, which was a fantastic unit, had approached the crossing using "mission command," which gave subordinate leaders maximum flexibility on how to accomplish their specific tasks. However, as I listened to their lessons learned, this highly prized method proved to be their downfall at a river crossing. Each subunit commander controlled their portion of the operation, and each unit attempted to maximize fires and forces at key points to "blow through" the enemy. Unfortunately, this created a huge bottleneck of forces on the bridgehead that trapped one another as vehicles were destroyed and the enemy reacted. We would have to be draconian with the limited terrain on the near side of the river to avoid the same fate.

My very young S3, Rais Sanchez, quickly grasped what I wanted based on what I had seen when I was a young platoon leader at Fort Polk. I felt I knew exactly what we needed to get to, but the entire weight of translating that to actionable tasks to a nascent unit in just forty-eight hours fell to her. She designed a massive staging grid controlled with an iron fist of MPs and tight control measures. Every action was captured on a large synchronization matrix using code words since we had to use unsecure radios to link in all our units. Her work was brilliant.

My NCO mentors of the past had taught me a great idea was useless if you could not effectively articulate it. I envisioned our attack as a SWAT raid on a house. The MPs would establish a cordon in the city and control all units flowing to the river, like patrol police sealing off the block. The Romanian infantry would then seize the near and far side like police surrounding the house. The engineer units would then deploy like a SWAT ram breaching a door, but in this case, they would breach the river with bridges. Then the Polish mechanized infantry would push across like door breachers in a raid. Once they were across, the Romanians and MPs would push across and begin to expand the bridgehead and finish the enemy.

Even though we had a good concept of how to attack, we had several big problems. First, we had no air support or artillery fire to suppress the enemy across the river. Second, we had no boats to allow the Romanian infantry to cross the river. Third, we were going to do this at night without lights for our first time, driving dozens of heavy vehicles full of kit-laden troops over deep waters. And fourth, the lights from the city where we were going to cross would light us up like a Christmas tree. I thought we would never be allowed to do something this risky in the US. Then it dawned on me that the responsibility was mine.

At twenty-four hours out we got simulated artillery support and actual attack helicopter support for the mission. 2CR sent their rubber rafts, and the Romanians boldly began to train on them. I even noticed that it appeared many of the streetlights had been turned off in the area, solving many of our spotlighting issues. I turned and asked Bila, my Romanian liaison, how he got permission to turn out the lights. He smiled and picked up a rock and threw it. He had personally thrown rocks to break the lights for us. You must love lieutenants, no matter what country they come from. We were as ready as we could be with just forty-eight hours to plan and execute.

As the sun went down on the night of the attack, it was surreal. We had vehicles staged in small groups spread out over twenty miles. We had used vegetation to camouflage them to deceive the enemy into believing we might cross the river in a rural location. However, hundreds of locals lined the streets curious to watch the show, making concealment nearly impossible as they shared everything they saw on social media. I could do nothing at this point except trust in the team and pray our dive teams would not be needed to rescue anyone.

It turned out to be a long, sleepless night with Sergeant Strass covering my six and Rais orchestrating the fight. My mobile command post was established just above the bridgehead, and I was able to watch the operation unfold. It was a very weird feeling to try to hold back from inserting myself into the tactical fight when not needed. I hated just waiting and watching, but Rais battle tracked the synchronization matrix beautifully, and leaders at every level adapted to the fight.

All four bridges in place and operational as the sun rises

The enemy, after initially falling back from the attack of the Romanians on rafts, circled to the flanks and began to apply pressure to squeeze the pocket. However, the German bridging unit was prepared and used their mobile bridge sections to ferry across several Romanian Humvees to provide heavy weapons support. This was followed shortly by the Polish mechanized company crossing and smashing the enemy reinforcements moving down to the river. Meanwhile, the MPs and engineers worked seamlessly to both control the flow and build the bridges under fire. When the sun came up, we were treated to the most beautiful sight of all four temporary bridges built and in operation. We had no serious casualties and only one accident, when an ambulance carrying simulated wounded misjudged a trail using night vision and rolled over on its side.

This accidental and ill-prepared exercise helped form my contextual philosophy for training the brigade. Hire good people, take care of them, and ask them to do tasks far greater than they think they can do while letting them know it is OK to fail. Failure is where growth and learning occur. My job was to create a climate and culture that supported it. Looking back, we did a lot of exercises and missions like our trip to Romania, from Poland to Hungary to Georgia. My top-rated leaders were not the ones who succeeded in the individual missions the most. They were the ones who created a positive climate and culture and took care of their people.

The day I gave up the Eighteenth MP Brigade colors, I knew I could never do anything in the army that would rival commanding such an amazing unit. These were the soldiers who protected this great country; they are our army. And they are out there today ready to defend it. I truly miss the great soldiers of the Ever Vigilant (EV) Brigade. They were truly Europe's edge.

CHAPTER 21:
COCAINE TAKES ITS TOLL (2019–2024)

❧

The Army is hard physically and mentally,
but anything worth it is.

I HAVE TOLD MANY audiences over the years that the army is like crack cocaine. You know it is bad for your body, but when the high wears off, you want another hit! Well, as I neared the end of my brigade command, the effects of the army were beginning to be too much for me and our family. I no longer thrived on high stress. I saw doctors almost weekly for dozens of systemic injuries and found it harder and harder to stay combat ready. At the same time, after fifteen moves Angela was tired of uprooting, and Hunter was looking at becoming a sophomore in his eighth different school system. The army wanted more, but there wasn't much more to give.

The MP Corps thought I was postured for the next rank. Not because I was that good, but I believe because my senior rater who said I was good was a rising general officer destined for his fourth star. I still didn't think I wanted to become a general, but the MP

Corps offered me an excellent chance to be the deputy comman-
dant of the regiment. This would allow my son to finish high school
in Missouri and my wife to enjoy some more time with me, and I
could give back to soldiers for a few more years without telling the
army no. I jumped at the chance. However, God had other plans.

As ironic as it sounds, because the army pulled me out of
Afghanistan two months shy of a year to take command of a unit
last minute, I was not given joint qualification. Joint qualification
was a requirement to make general officer, and the position I was
in in Afghanistan was designed to qualify me. I was in the process
of submitting my records and missions in early 2019 that I did in
brigade command to get the last few credits I needed when the rug
was pulled out from under us. The then regimental commander,
who had personally asked me to come be his deputy, reneged on
his word. He was too scared that I would not get joint qualification.
He told me I needed to do another joint assignment to ensure I got
qualified. I bluntly told him that he was breaking his word to me,
and he simply said he could not risk it. It was the last time I ever
spoke to him.

The MP Corps lieutenant colonel who managed colonels, including me, sealed my decision. I cannot remember his name, but
I remember thinking he was an idiot. He told me he was going to
assign me to the joint staff in the Pentagon, likely as a three- or four-
star executive officer to ensure I got joint qualified and known in
the senior community. I still was in shock that lacking sixty days in
Afghanistan was the difference as to whether I was ready for general
officer. So I told him I would not put my family back in the DC area.
He told me he would then put me on orders for another assignment
where I would deploy again, and when I refused, he said he would
put me out of the army in ninety days. I told him, "Someday you
and I will meet." I was just finishing four of the most demanding
years of my career, and here was this clown telling me if I didn't do
what he wanted, I was done. I thought, *hold my beer.*

Luckily for me, the army has twenty times more good leaders than bad. Both my boss, Major General Shapiro, and Lieutenant General Cavoli took good care of me. They made sure I knew that if I said no to the joint assignment, I was taking myself out of the running for the next rank. Once they understood my determination, they asked where I wanted to go before I retired. I asked to go back to Alabama, largely so my son could finish high school where they still prayed before each football game and no one got offended. General Cavoli blew through every roadblock my branch presented, and within two weeks I had orders to go to the Air War College in Montgomery, Alabama, to teach newly minted colonels.

I must admit I felt totally disloyal to the army, because in almost thirty years, I had never said no to what it asked of me. I became irrelevant overnight to the Military Police Corps and struggled with losing the high from the constant stress in the army day-to-day grind. However, over time the assignment to Maxwell became my saving grace.

As I finish this chapter, I am in the final months of finishing my thirty-four years in uniform. I arrived at Air War College to teach in 2019, expecting to retire when my son graduated high school in 2022. However, he decided to go to Auburn University on an army scholarship, and we decided to retire here, so I was allowed to remain here until retirement in 2024. Teaching senior officers has been a way for me to give back while I have had time to deal with both my mental and physical scars from a fantastic career.

My experience at Air War College has helped me greatly to be prepared to leave the army. In many ways the air force is closer to a civilian organization than a military one, or at least the army that I grew up in. They work normal eight-hour days including physical training, avoid risk by moving methodically, and prioritize individual desires and job satisfaction highly, and enjoy many creature comforts foreign to the army and its warrior mentality. At first I just made fun of their perceived weakness. However, the longer I was exposed to it, the more I realized it breeds a very happy and knowledgeable cohort that makes the air force such a great partner. It just could not work in the realm of ground warfare.

Over the last four years, I have taught several hundred senior officers as they begin their journey into the senior levels of military service. I learned that to stay sharp at any age, you must continue to challenge yourself, continue to learn, and resist the urge to simply look back. Growth comes in the zone of uncomfortableness, so I am attempting to not be comfortable and looking forward to finally hanging up my uniform.

As my final duty in uniform to my family, I am finishing this book. My son, Hunter, asked me to share some of my crazy stories and lessons I've learned, and I hope I did it justice. In the end I would not change my journey, and I am super proud my son is becoming a soldier too. If you have never served, you might find yourself thinking this is either fiction or the army is crazy. But I bet if you have served, you have flashed back to some of your crazy adventures and said to yourself, *If you think that is bad, I remember when, no shit, there I was…*

HEROES

THERE ARE LITERALLY THOUSANDS of heroes without whose help I would not have made this incredible journey. I cannot list them all, but below are just a few to whom I owe so much because of their mentorship, friendship, and example. I want to acknowledge publicly that they are my heroes:

Maria Thompson – More than a teacher—a beacon of hope

Donald L. Chumley – Saw something more in me and motivated me to go after it

The Linville family – Loved me unconditionally

Tina Harkey – My absolute best friend in college

Larry and Wesley Bauguess – Made me reflect and be a better human

Chris Henshaw – The first MP captain I wanted to be

My Ranger Buddy Rulla – I owe my Ranger tab to him

Mountain Phase RI who "hated MPs" – Kept me from quitting and may never know

Mark Boucher – Taught me the ways of the old breed

Anthony and Donna Adams – First to show me what an Army family should be

Niave and Andy Knell – Showed me how to be passionate and inspire those around you

Chip Balk – Taught me to work hard and have fun at the same time

Garret Fawaz – Demonstrated extreme patience with a young lieutenant

Ted and R'ami Spain – The first senior leader team I would go to hell and back for

John Narcisse – Taught me what a professional NCO should be

Carl Lally – Kept me humble while always backing my play

Stephen Rodriguez – Showed me respect is not about rank, but about character

Tim and Susan Weathersbee – Showed me how to care for troops as an officer

Jeff Stewert – Taught me how to be a commander

Tim and Kelly Skinner – Showed Angela and me how to be a successful couple in the military

Jody Lupo – Showed me those who follow us are often smarter than we are

Janette Kautzman – Taught me courage is contagious, and she was fearless

Gary Pauley – Demonstrated daily that fun is what we are charged to create

Rex Sprague – Exemplified courage and character doesn't end with your service

Jennifer Gray – Taught me to think beyond my small portion of the fight

Kevin Schnieder – Showed me holding the "standard" transcends rank and time

Jeff "Toz" Toczylowski – Taught me to love and have FUN daily, because life is too short

Jeff and Les Sweeney – Were super close battle buddies to Angela and me

Andre Kendrix – Showed me that it is not what happens, it is what you do about it that matters

Steve Bracero – Showed me that the spirit of a true NCO is the most powerful weapon

Chris Heberer – My ride or die

Mike Thompson – My ride or die

Dave Koonce – Showed me you don't have to be loud, to make a huge difference

Dianne Manning – Loved me like a mother

Pete Keough – Taught me the power of prayer in combat is beyond any bullet

Carolyn Bronson – Showed me courage has nothing to do with rank

Mark and Mary Green – True lifelong battle buddies to Angela and me

Brad Graul – Taught me that God puts us where we need to be, regardless of whether we want to be there

Steve Basso – Proved that iron sharpens iron

Dave Sage – Showed me that doing the right thing is always the right army answer

The Cremer Family – Simply our family

Daniel Bolger – Showed me senior leaders must live to a higher standard

Richard Walters – Taught me if you are going to do it, win

David Flynn – Taught me that leaders never stop learning

Ricky Trull – Showed me God's power and a father's love

Bobby Lungrin – Taught me that men of character are priceless

Travis and Becky Jacobs – The definition of character and class as a family team

Rodney and Fredia Johnson – Showed us that "faith and family" are second to none

Michelle Ambul – Showed me the spirit of the next generation is as strong or stronger than ours

Lee Walters – Showed me the more senior you become, the humbler you must be

Mark Avril – Taught me that senior leaders must know which details to master

Duane Miller – Taught me it is OK to be passionate—just be careful leading with it

Steve Shapiro – Taught me that leading at the senior level is all about trust—give it to get it

Rais Sanchez – Fearlessly translated my "good" ideas into great and reasonable ideas

Patricia Straszewska – Showed me that hard work and moral character are combat multipliers

Judd Young – Taught me your subordinates often should be your boss

Ted Pearson - Showed me that senior-level NCOs are true warrior scholars.

My entire Eighteenth MP BDE medical team – Constantly pieced me back together and kept me in the fight

David Glaser – Showed me grace when mentally and physically I was broken

Bill Lewis – Coached me tirelessly on life after service

Jonathan Hart – Mentored our family and furthered our journey with the Lord

Angela Schmick – My saint, best friend, and one true love of my life

Hunter Schmick – My pride and joy, my battle, and our next generation of warrior

www.ingramcontent.com/pod-product-compliance
Lightning Source LLC
Chambersburg PA
CBHW022125080426

42734CB00006B/239